PENGUIN BOOKS

CHOOSE TO LIVE

Joseph D. Weissman, M.D., is clinical assistant professor of medicine at the UCLA School of Medicine and director of the Medical Center for Health and Longevity in Torrance, California. He specializes in preventive medicine, immunology, and the treatment of allergies. In addition to his many professional activities, he is an avid long-distance runner who has completed three ultra-marathons (45-50 miles) and fifty-five marathons.

Choose to Live

JOSEPH D. WEISSMAN, M.D.

PENGUIN BOOKS

PENGUIN BOOKS
Published by the Penguin Group
Viking Penguin Inc., 40 West 23rd Street,
New York, New York 10010, U.S.A.
Penguin Books Ltd, 27 Wrights Lane,
London W8 5TZ, England
Penguin Books Australia Ltd, Ringwood,
Victoria, Australia
Penguin Books Canada Ltd, 2801 John Street,
Markham, Ontario, Canada L3R 1B4
Penguin Books (N.Z.) Ltd, 182–190 Wairau Road,
Auckland 10, New Zealand

Penguin Books Ltd, Registered Offices:
Harmondsworth, Middlesex, England

First published in the United States of America by
Grove Press, a division of Wheatland Corporation 1988
Published in Penguin Books 1989

10 9 8 7 6 5 4 3 2 1

LIBRARY OF CONGRESS CATALOGING IN PUBLICATION DATA
Weissman, Joseph D., 1930–
Choose to live/ Joseph D. Weissman.
p. cm.
Reprint. Originally published: New York: Grove Press, 1988.
Bibliography: p.
Includes index.
ISBN 0 14 01.1584 6
1. Environmentally induced diseases—Prevention. 2. Environmental
health. 3. Pollution. I. Title.
[RB152.W46 1989]
616.9′8—dc19   88–31672

Printed in the United States of America
Set in Times Roman
Designed by Irving Perkins Associates

To my wife, Phyllis, my children, Jill L. Tate, Laurie Jean Bucher, and Captain John R. Bucher, U.S. Marine Corps, and my lovely grandchildren, Michelle "Mikki" Nicole Tate and George Parker "Parkey" Tate.

Contents

Part III   THE RESEARCH

Acknowledgments

George Edwin Tate, Jr., my son-in-law and co-founder of a leading computer software company, was just beginning to enjoy amazing business success when a heart attack cut his life short. George's untimely death stimulated the research for this book, which changed my perception of medicine and my own lifestyle. I hope the book will contribute to saving many lives.

Writing a book is like coaching a team or leading an orchestra; the writer, coach, or conductor cannot do a creditable job without the help of many people. It is impossible to mention here all the friends, colleagues, and associates who made contributions to this book. But I do especially wish to extend my gratitude and thanks to Electra Brooks, who first suggested that I write the book; James Neyland, my editor, whose superb writing skills made the book more readable; Harold Roth, my literary agent, who was adept at opening previously closed doors; and Fred Jordan and Joy Johannessen at Grove Press, who made some very astute suggestions for improving the book's final form, and who made this all possible.

I also wish to thank Bernard Jaroslow, Ph.D., Stuart Wong, M.D., Spencer H. MacCallum, Alvin Lowi, Jr., Benjamin Rosin, M.D., Gildon Beall, M.D., Douglas Heiner, M.D., Robert Morin, M.D., and numerous other associates for reviewing parts of the manuscript and providing helpful criticism and suggestions; Deanne Hockett for her help with graphs and charts; Anita Klecker, chief librarian of the Torrance Memorial

Hospital Medical Center, for her assistance with research; my staff and patients for their patience; and finally, those twentieth-century inventions that simplified my life, the Compaq Deskpro Computer, MultiMate Advantage word processor, and dBase III Plus database software for my charts and tables.

Introduction

In modern technological societies, we are faced each day with numerous choices. The choices we make may be small or large. We choose our work or careers; we choose whom we will marry and when we will have children; we choose where we will live and how we will spend our leisure time. The list is virtually endless. In most cases, we are aware that our freedom to choose also implies some degree of responsibility. Our choices and their accompanying responsibilities are second nature to us. We do not need to be reminded of the consequences if we choose unwisely.

However, we face our most important personal choice armed with so little knowledge that we may make it unwisely and perhaps irresponsibly, without understanding the full consequences. It is a choice made every time we light up a cigarette, take a drink of tap water, or order a fast-food burger. Those selections can determine whether we will be energetic, alive, healthy, and vibrant in our daily lives, and whether we will reach a ripe old age with a good quality of life, avoiding the killer diseases that have become so prevalent in our era—cancer, heart disease, arthritis, Alzheimer's, multiple sclerosis, and diverticulosis, among others.

Most people are unaware that they can choose a life of excellent health, remaining active, trim, and alert. They assume instead that ailments arise from causes beyond their control—from fate or genetics or the bad luck of encountering a virus.

Or they do not really believe that the serious illnesses will touch them, rationalizing that they are somehow protected by a merciful God, that they have always had good luck, that because their parents and grandparents lived long and healthy lives, they too will be untouched. They believe they can live the high-risk modern lifestyle, eating whatever they want, doing whatever they choose, and expect to feel healthy indefinitely.

However, there are strong indications that in addition to genetics, choices of diet and lifestyle in our industrial societies plays a very large part—perhaps the largest—in whether we will remain vibrant past our prime or fall prey to various ailments. My research reveals that many diseases have developed within the last 200 years as probable by-products of the Industrial Revolution. The very technology we have created to make our lives easier and to rid ourselves of disease is now attacking us through toxic chemicals that have been introduced into our environment and food supply both intentionally and accidentally, without our realizing their full cumulative effects. There is virtually no soil or water supply in the developed world that remains unpolluted by these toxins, and consequently they are also in almost all the food we eat, the water we drink, and the air we breathe.

How, then, can we choose to avoid these poisons and live active, vibrant, longer, and healthier lives? It is impractical to consider moving millions of people to a tropical island or a remote jungle area or the outback of Australia, places still relatively untouched by progress. Nor is the solution to rid ourselves of our technology and industry; we are now dependent upon the marvels we have created. In the long run, for the sake of our children and grandchildren, we must find ways to have our technology without the toxins, but that will surely take longer than our own lifetimes to accomplish. Since it took more than 200 years to accumulate the current extensive pollution, it may take as long, if not longer, to get rid of it.

Our own choice—if we choose to live—is a more personal one, and one that must be made with full responsibility for our own bodily health. If we make this choice, we face numerous other choices, some of which may be exceedingly difficult for us. We must become involved in decisions about what we eat and drink and about the safety of our environment.

Some toxins have become so pervasive that it is no easy task to avoid them. It can be done, but it will require following a carefully planned program of diet and lifestyle changes, facing each day with an awareness of the choices needed to maintain health.

I have developed a ten-point program for decreasing exposure to the toxins that interfere with our lives. I am not offering quick-fix gimmicks, easy panaceas, or a temporary weight-loss diet. What I propose is a different, perhaps an austere regimen, but if we wish to live long and healthy lives, it is important to put some effort into it. Some people may find some of the points impossible to adhere to; that is understandable. However, the program is intended as advice or recommendations that will enable individuals to take responsibility for their own lives, making educated choices to maximize their health and minimize risks and dangers.

To help the average person adjust to the major changes in diet and lifestyle presented by the program, I have devised a gradual, step-by-step ten-week procedure in which only one or two major alterations are introduced each week. I include specific suggestions to make the changes as easy as possible.

Most people have been aware for some time that obesity, use of tobacco and alcohol, and lack of proper exercise can be harmful to health. Much less familiar are the dangers related to animal food sources, drinking water, beverages, and processed foods, all of which contain toxic chemicals associated with technological society. These toxins and their actions within our bodies are what we must avoid to remain healthy. I believe that they cause many of the diseases that are so prevalent in the twentieth century.

The diet I recommend is essentially a vegetarian one, but I prefer to call it a low-toxin diet since there would be no health-related objection to animal food sources from an unpolluted natural environment. For those who will simply not accept the complete low-toxin diet, a compromise that allows occasional meats or other animal products may still result in an improvement in health. Some benefits are possible even if you choose to follow only a few points of the program.

There are, of course, standard arguments against vegetarianism. Isn't *Homo sapiens* an omnivorous animal? Yes, we adapted to this pattern early in evolutionary prehistory, like most primates. The foods available to early humans varied widely, but they could readily adapt to their environment and the available food sources. However, during the Industrial Revolution we altered nature, putting into the environment chemicals that have changed the constitution of our food sources. The concentration of these chemicals in animal sources is considerably higher than in plant sources, in some cases dangerously high, for toxins accumulate and are readily stored in cholesterol and fatty tissue.

The argument here for a low-toxin diet is not philosophical but pragmatic; it is purely a determination from extensive research. It is a recommendation for achieving good health and longevity—and avoiding many modern diseases.

Many people are wary of vegetarian diets because they have heard that protein and certain other nutrients cannot be obtained from any food source but meat. It is important to dispel this myth. The fact is that our protein requirements are considerably lower than most people realize; the Western diet contains far more protein than is essential for the body's needs, perhaps even dangerous amounts. Some reports indicate that Americans eat three or more times the amount of protein they actually need.

Amino acids are the building blocks of protein. There are nine essential amino acids the body is unable to manufacture (histidine, isoleucine, leucine, lysine, methionine, phenylalanine, threonine, tryptophan, and valine). We must obtain these from food, but not necessarily from meat. In fact, we are able to use the amino acids obtainable from vegetable sources very efficiently. It is virtually impossible to develop a protein deficiency on a vegetarian diet. Vegetables, legumes, grains, nuts, and seeds contain more than enough protein for the body's growth and maintenance.

Another misconception is that it is very difficult to obtain *complete* proteins on a vegetarian diet. The erroneous concept that plant protein is inferior to meat protein stems from research on rats done by Osborne and Mendel in 1914, and it has been fostered by the meat and dairy industry. This research also underlies the idea of mixing and matching plant foods at a single meal (e.g., rice and beans, corn and peas) so that their amino acids will combine to form complete proteins. Such mixing and matching does seem to be important for rats fed nothing but artificially purified amino acids in laboratory experiments, but there is no evidence that humans need to eat mixed vegetable foods at the same meal or even on the same day; simply eating a variety of plant foods will supply all the essential amino acids and provide for complete proteins. To see how silly the concept of mixing and matching is, one need only look at primitive vegetarian peoples who have never heard of it and remain perfectly healthy.

There is no reason to be concerned about nutritional deficiencies on a vegetarian program. Vegetables provide plentiful supplies of protein, calcium, iron, ascorbic acid, and all the other vitamins and minerals our bodies need. The Seventh-Day Adventists, many of them total vegetarians, have followed their own form of the low-toxin diet and have an outstanding health history. Compared to the omnivorous population, they have an extremely low incidence of cancer, heart attacks, and many of the other diseases of civilization. A reluctance to change to a vegetarian diet may have social and psychological causes, but it has no nutritional foundation.

Some people may simply feel that they cannot give up their favorite meat or fish. In that case, I suggest that animal food, instead of being part of the daily diet, become an occasional treat. This seems a reasonable trade-off for healthy living when your favorite repast may have become a carrier of slow-acting poisons. If you value life sufficiently to follow the low-toxin program, you'll find that after a while your tastes will change and you will become less dependent on your old favorites. And since this is not just a weight-loss program but a program for supreme health as well as survival, you can modify the diet on occasion, knowing that what matters is your overall ability to stick to it. It is *your* body, *your* life, and it is *your* responsibility to take care of it. You are always free to choose.

What about eating out? Won't that be impossible? Not impossible, but you will have to choose restaurants wisely and select carefully from the menu.

But isn't the low-toxin program expensive? Surprisingly, the program will actually save you money. Your food bill will drop by as much as 30% or 40% since you will have omitted the most expensive items—meats and dairy products. Considering that it takes a minimum of 16 pounds of grain to produce 1 pound of beef, you can easily see where the bulk of the savings occurs. Also, most processed foods and junk foods require a great deal of advertising; if you stick to "natural" foods, this advertising cost does not become part of your food expense.

Even using distilled water and taking vitamin supplements daily, as the program recommends, should not add a terrible burden to your budget. You will save far more with your dietary changes than you will spend on these items. If you follow the program correctly, it should also decrease your need for drugs and medical care. This can certainly result in considerable savings.

Finally, saving all the money in the world is unimportant compared to the good health you can enjoy on this program.

Before beginning the program, you will naturally want to have a fuller understanding of why it is necessary, and that information is supplied in Part I, "A New Approach to Health and Longevity." In Part II, "The Program," I discuss each of the ten points at length, stressing specific dos and don'ts in the battle for fitness and health (you may be particularly interested in Chapter 14, which includes case studies showing what the low-toxin program has done for specific individuals with a variety of health problems). Part III, "The Research," presents the material on which the conclusions and recommendations in the first two parts are based. Throughout I have kept the discussion as nontechnical as possible. Though this book was written for a general audience, physicians should take the time to read it and especially to consider the evidence contained in Part III.

What I present here is new, but the historical data and the scientific facts on which I base my approach are accurate and verifiable. Moreover, the low-toxin program has become an important part of my medical practice and has improved the quality of life for many of my patients suffering from a variety of conditions. In my own case, it has also been remarkably effective. Though my conclusions await final scientific validation, I have no doubt that future medical research will confirm them. Until then, we can choose to save our lives on our own responsibility. For, to paraphrase Hippocrates, life is short and science is slow.

Part I

A New Approach to Health and Longevity

Chapter 1

The X Factor

During our era, a number of myths have grown up around the subjects of disease, health, and longevity. Though many of those myths are only popular misconceptions, some are also an accepted part of medical belief. True, we have made great strides in medicine and science in the twentieth century; there is no denying that. But to make further meaningful progress we shall have to get rid of some of the mistaken ideas that stand in the way of a full understanding of what has been and remains to be achieved.

MYTHS OF HEALTH, DISEASE, AND MEDICINE

The most serious myth, one accepted by medical science and the public alike, is that most of the major noninfectious diseases of the twentieth century have always been with us, simply a part of the human condition from the very beginning, and only brought to light in recent times because of improved diagnosis and greater life expectancy.

Closely linked to that myth is another—that many of these diseases are "degenerative." Fatalistically, we have come to accept that cancer, coronary heart disease, Alzheimer's disease, arthritis, diverticulosis, diabetes, and other ailments are to be expected with advancing years.

But both assumptions are patently untrue and inadvertently self-serving, like the related myth that we are living longer and healthier lives today as a result of scientific advancements. That one clearly ignores or overlooks facts of history and encourages the general public's confidence in medical science.

Consider these facts. One hundred years ago

1. *coronary artery disease* (also called coronary heart disease, or CHD) was virtually unknown throughout the world. The first description of coronary occlusion (blockage associated with coronary artery disease) and "heart attack" appeared in the medical literature in 1910. Today, coronary artery disease is the leading cause of death.

2. *cancer* caused approximately 3.4% of all deaths in Europe, and less in America. A century earlier, it was responsible for fewer than 1%. Now cancer is the second leading contributor to death, claiming 1 out of every 4 men and 1 out of every 5 women. (See Table 17.3.)

3. *diabetes* was extremely rare; 2 out of every 100,000 Americans had the disease, compared to 1 in every 20 today (more than 12 million Americans are afflicted). Diabetes and its complications are the third most common cause of death.

4. *Alzheimer's disease* did not exist. The condition was first recognized in 1907 by the German physician Alois Alzheimer. Presumably due to its rarity until recently, it has not yet been included in the U.S. vital statistics tables. Over 3 million Americans have this condition, which is considered the fourth leading cause of death.

If these facts are a surprise to you, consider that I am a physician and should have long been aware of them—but I wasn't. The information was shocking; I discovered that the billions spent on research, newer diagnostic techniques, organ transplants, coronary bypass procedures, chemotherapy, radiation, and various drugs have not appreciably altered the advance of the killer diseases.

We have been led to believe that the incidence of many of these killer diseases is a function of an aging population and is nothing more than a result of bodies growing older, wearing out, and falling apart. However, this does not account for the number of younger people who suffer from Crohn's disease, arthritis, cancer, heart attacks, multiple sclerosis, and other ailments that have been on the rise in the twentieth century.

Where have all these conditions come from? Certainly not from the aging process alone. Young people today acquire them, and older people in previous centuries didn't. Today even newborns and very young children can be victims of cancer and leukemia. This was unheard of at the turn of the century. Moreover, there is reason to believe that seemingly unrelated conditions affecting children and young adults—congenital birth defects, sterility, endometriosis, premenstrual syndrome, Down's syndrome, behavior and learning disorders, to name a few— may have a common bond with the killer diseases. In short, all the evidence suggests that in relatively recent times a great number of diseases have been on the rise, undeterred by the advances of modern medicine.

A bright note in the progress of medicine has been in the area of infectious diseases, where we have undeniably made great strides. Vaccines, serums, and antibiotics have been in-strumental in saving many lives. We in medicine can be justly proud of this achievement. However, our pride must be tem-pered by the knowledge that infectious disease began to decline slowly but significantly at the beginning of the nineteenth cen-tury, a full century and a half prior to the introduction of most of these modern miracles. Paradoxically, this decline was in-timately associated with the rise of the "degenerative" diseases, and both trends may have been caused by the very same factors.

Throughout history, until the modern era, the major causes of death were infections; tuberculosis was almost always num-ber one (see Table 16.7). Medical science did not know what caused tuberculosis. Various incorrect theories blamed night air, swamp gas, and overwork, among other factors. Sanatoriums offered the water cure and other ineffectual treatments; most often the victims confined to these institutions did not return alive.

It is difficult today to appreciate how little help medical science provided during the nineteenth century and earlier. There were no antibiotics and almost no vaccines; available treatments were primitive, torturous, and often dangerous. Contracting a major infection frequently resulted in death, either from the disease or from the treatment. Lucky were those who did survive; infants and toddlers almost never did.

In the last half of the nineteenth century, progress slowly resulted from the recognition that microorganisms could be responsible for infections. Joseph Lister, John Snow, Robert Koch, Louis Pasteur, Ignaz Semmelweis, and others made contributions that were responsible for the gradual introduction of sanitation. We generally associate the decline of infections with vaccines and antibiotics, but with the exception of a few vaccines, these were not developed and introduced until very recently, in the era commencing with World War II.

During the last century, when infections were still rampant, conditions such as Parkinson's disease, systemic lupus erythematosus, multiple sclerosis, amyotrophic lateral sclerosis (Lou Gehrig's disease), muscular dystrophy, ulcerative colitis, and many others were gradually emerging (see Table 16.1). Their development coincided with progressing industrialization. Some contend that these diseases have always existed, but the evidence to support this is meager at best. On the contrary, what we are seeing are historically new maladies.

It is difficult to understand how people, physicians in particular, can believe the myth that the so-called degenerative diseases have been around forever but have only recently been discovered because of improved diagnostic skills. This belief does not conform with the facts. Many physicians in earlier centuries were expert diagnosticians who used simple but effective techniques that are now a lost art. Their ability to diagnose and describe both rare and common infections solely through the senses of smell, taste, touch, sight, and hearing is well known. And they were able to do this for centuries prior to the birth of the germ theory in the 1880s, without benefit of knowing the cause of the conditions. They were also able to recognize and diagnose a few noninfectious diseases—gout, diabetes, cancer, and perhaps myasthenia gravis. It hardly seems likely that these excellent observers would have been incapable of recognizing and describing other major modern-day diseases, particularly in the final stages when diagnosis is relatively easy. We must assume for lack of descriptive evidence that the killer diseases were absent or rare.

It will be necessary to dispel another myth—that our ancestors did not suffer from the "degenerative" diseases because they died young of infections and therefore did not live long enough to acquire them. In virtually every grade school health class, we are told that because of the miracles of modern medicine, we can expect a lifespan considerably longer than our ancestors enjoyed.

The real explanation for the apparently longer modern lifespan lies in statistical averages. Mortality rates among infants and children in earlier centuries were high because of infectious diseases. Those who died young, particularly those who died in infancy, brought the *average* lifespan down considerably. However, many people lived to ripe old age once they survived their youth (see Table 1.1). If, for example, you had reached the age of 45 back in 1849, the year of the Gold Rush in California, your life expectancy would hardly have been different from that of a 45-year-old today; now you might have an advantage of 3 to 5 years, and that is all.

TABLE 1.1
Number of Persons Age 50 or Older in England and Wales

Year	Total Population	# 50 or Older
1821	12,000,000	1,486,400
1851	17,990,000	2,547,200
1901	32,621,263	4,790,000
1951	43,758,000	12,216,300

Source: Joseph Whitaker, *Whitaker's Almanac* (London: William Clowes, 1984).

It is life expectancy from *birth* that has changed most dramatically, and that because of our improved chances of surviving infancy. Babies born today can expect to live 25 years longer than infants did 100 years ago. In 1850 the average life expectancy from birth was 40 years; it rose to 50 by 1900; it is approximately 76 today. But this is only a *statistical average*. It is not the maximum possible lifespan.

Maximum lifespan is the greatest age obtainable by a species. In humans, this maximum has been 110 to 120 years and has not changed appreciably in recent times. In Europe 15,000 years ago, average life expectancy was about 28 years, but maximum lifespan was not much different among the people of the Roman Empire from what it is today in the United States.[1]

While it is true that medical science has made enormous advances in combating and defeating infectious diseases, it does not appear that these achievements occurred early enough to deserve more than minimal credit for increased life expectancy. As mentioned, the great breakthroughs in medicine did not come until the end of the nineteenth century and during the twentieth; in fact, the use of sulfa, antibiotics, tuberculosis drugs, and many vaccines did not become standard until the 1940s. Yet the incidence of most of the diseases that once cut life short in infancy or early childhood—such diseases as tuberculosis, whooping cough, scarlet fever, typhoid, measles, diphtheria, cholera, and infant diarrhea—has been declining steadily since the early 1800s (see Tables 16.2–7). Life expectancy throughout the industrial world began to rise even before 1800, at a time when medical treatment consisted of such primitive techniques as cupping, leeching,[2] blistering, bloodletting, enemas, purging, and water cures.

Some attribute the decline of infectious disease and the ex-

---

[1]Life extension beyond previous maximums has been accomplished in laboratory animals by decreasing their food intake or lowering their body temperature. There is a possibility that the maximum age barrier among humans can be extended by decreasing food intake.

[2]At the height of the bloodletting craze in 1830, 20 million leeches a year were used in France.

tension of the lifespan to better living conditions—improved housing, sanitation, nutrition, etc. However, the crowded slums of the industrial cities of the early nineteenth century can hardly be called better housing; crowding is known to hasten the spread of infections, not to contain them. Sanitation was not considered important and was not seriously practiced until rather late in the century. Adulteration of food was rampant, milk was polluted, raw sewage was permitted to flow into rivers that supplied drinking water. Although nutrition was improving, good nutrition was entirely the province of the rich; most of the poorer slum dwellers suffered from malnutrition. Some of the worst famines in history occurred in the early nineteenth century.

It is more than mere coincidence that the infectious diseases began their decline in the early years of the Industrial Revolution. The very same factors that eventually led to the "degenerative" diseases—the chemicals and toxic wastes produced by industrial society and unknowingly incorporated into the food, air, and water—may have had for their earliest victims some of the bacteria and viruses that were the longtime enemies of mankind.

In this century, sanitation, the chlorination of water, and the pasteurization of milk must be given further credit for the elimination of childhood intestinal diseases and for rising average life expectancy. However, since these measures were not regularly applied until the end of the nineteenth and the beginning of the twentieth century, they alone cannot account for the gradual rise in longevity rates that began a hundred years earlier.[3]

The Industrial Revolution and its wastes unquestionably played a role both in extending life expectancy and in creating what can aptly be called the new man-made diseases.

---

[3] Although chlorine may have been having unintended effects, both good and bad, even before the chlorination of water, since it was a waste product discarded into streams and rivers in the early nineteenth century.

THE "X FACTOR" IN DISEASE

During the last 200 hundred years the world has undergone a unique period of rapid industrialization. The Industrial Revolution brought with it new man-made chemicals: chlorine and its compounds, coal-tar derivatives, pharmaceuticals, petrochemicals, etc. Steam and electrical power, the internal-combustion engine, and the mass-production of consumer and industrial goods made their debut.

All industry, past and present, creates by-products and wastes that require disposal. The only means of eliminating them are burning (with the subsequent development of toxic smoke), disposal in nearby waterways, or burial. The emergence of industrialization, with its production of masses of waste, coincided with the discovery—and presumably the first appearance—of many new diseases.

Underground waste disposal ultimately intrudes upon water aquifers, and burning wastes pollutes the air; thus toxic materials are deposited in farming areas and finally make their way into water and food supplies. All foods are affected: fruits, vegetables, grain, fish, poultry, meat, eggs, dairy products. Some foods store more toxins than others, for some are bioconcentrators and biomagnifiers. Generally, *all* animals are bioconcentators—from fish, mollusks, and birds to cattle, sheep, and humans. The absorption and retention of poisons in animals is far greater than in plants. The greatest concentrations of toxins occur in animal fat and cholesterol, for many chemical toxins are fat-soluble; muscle tissue, eggs, and milk are not exempt, however. (See Table 1.2.)

The various toxic substances in the environment constitute the X factor. This term derives from the Greek *xenobiotics*, a word research scientists use to describe substances foreign and harmful to living creatures, including man. Though some xenobiotics occur naturally in the environment (for example, ultraviolet irradiation from the sun, certain chemicals found in plants or created by volcanic activity), by far the greater number owe their existence to human intervention; these include man-made

TABLE 1.2
Sources of Pesticide Residues in the U.S. Diet

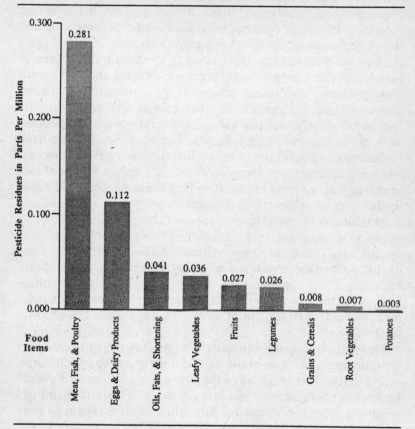

Source: Modified from data in G. Q. Lipscomb and R. E. Duggan, "Dietary Intake of Pesticide Chemicals in the U.S.," *Pesticides Monitoring Journal* 2 (1969): 162–69; and P. E. Cornelliussen, "Pesticide Residues in Total Diet Samples," *Pesticides Monitoring Journal* 2: (1969) 140–52, 5: (1972) 313–30.

poisons, pollutants, reactive chemicals, free radicals,[4] radio-active substances, heavy metals, and most pharmaceuticals and chemical food additives. It is the intrusion of the X factor that has been the major cause of both the decline of infections and the appearance of the new man-made diseases.

Most medical theories that attempt to explain the causes of various diseases do not adequately account for the absence of these diseases historically and in existing primitive societies. Our ancestors, like present-day primitive people, were virtually free of "degenerative" diseases. Even meat-consuming peoples such as the Eskimo, Pigmy, Masai, Samburu, and Navajo have had no multiple sclerosis or lupus, little coronary artery disease, and very little cancer. Heredity does not explain it, since their relatives who migrate to industrialized areas of the world begin to develop "diseases of civilization."

In addition to exercise, one crucial factor appears to explain the lack of "degenerative" diseases in these groups: they have, or did have, relatively pure sources of food and water. They drank water that was not altered by chlorine or other disinfectants. Their food sources—cattle, camels, birds, or fish—had access to pure water and food untainted by industrial smoke and pesticide residues. In other words, they were not exposed to the X factor.

From this brief survey, a pattern begins to emerge. The major noninfectious diseases of our era do not appear to be a natural, unavoidable part of life or of the aging process. And they seem to be relatively new or modern diseases. Those that did not originate after the Industrial Revolution at least seem to have been rare before this period. Almost all began very slowly at the beginning of the nineteenth century and seemed firmly established by the early twentieth century, many starting a dramatic rise thereafter as the traditional infectious diseases declined.

[4]Free radicals are very unstable, highly reactive molecules with odd numbers of electrons. They are created from stable compounds (and xenobiotics) by heat, radiation, chemical reactions, sonic vibration, and other means. Free radicals can be destructive to DNA, RNA, and tissue membranes; they are carcinogenic and are probably responsible for other diseases.

It must be noted that we now seem to be witnessing a reversal in the pattern of decline of infectious disease. We know that bacteria are capable of developing strains resistant to our most potent antibiotics. Insects, carriers of disease, also develop hardier offspring through mutation, their answer to our ever more exotic poisons. Today, as if out of nowhere, new viruses are emerging to defy the historic trend and to cause a variety of very serious illnesses: AIDS, Epstein-Barr syndrome, hepatitis, viral leukemia, cytomegalovirus (CMV), and others. As far as we know, their viral ancestors did not cause disease in earlier days. I believe that the recent resurgence of infectious disease is due to the same factors responsible for its historic decline: the chemical and other toxic wastes that have been saturating our environment since the Industrial Revolution. It is more than possible that these pollutants have created all the right conditions for pathogens to adapt and reproduce in increasingly virulent forms while the human immune system falters under their combined assault. Even tuberculosis is on the rise again, reversing its centuries-long decline.

The clear historical association between the accumulation of toxins in the environment and the emergence or recession of various diseases is not a pattern to be ignored in searching for the causes of—and the solutions to—these diseases. Once we link the ills of mankind to xenobiotics, the implications are enormous. The evidence suggests that we as a species are responsible for many of our diseases and that we must adjust our attitudes accordingly and adopt new methods of dealing with them.

Chapter 2

Combating the X Factor

While we wait for society as a whole to behave responsibly and repair the ecological damage we have done, it is important for individuals to take responsibility for themselves and to avoid drinking, eating, or breathing the chemical poisons that threaten them. Because it is impossible to avoid all xenobiotics, other steps, such as taking vitamin and mineral supplements and giving the body proper exercise, are also necessary to fight or counteract the dangers.

Of course, the average person cannot be expected to recognize all the enemies from the generalized picture presented so far. There are far too many threatening villains, with names even more complicated than those of the prehistoric beasts faced by our primitive ancestors. The ten-point low-toxin program is a simplified guideline for health and longevity. It will enable the average person to cope effectively with the risks and dangers without having to memorize long lists of technical names. Just as it was not necessary for our ancestors to recognize the frightening beast as a *Tyrannosaurus rex*, we do not have to know the term *butylcarbamoy-benzimidazole carbamate* to be aware that a household fungicide may be harmful to us.

Parts of the dietary program may sound familiar, for some

of my recommendations to avoid or concentrate on certain foods are consistent with other nutritional programs. But the reasons for the recommendations here may be considerably different from standard nutritional advice. Other points of the program are diametrically opposed to currently accepted nutritional standards. However, the objective is always to get all the essential nutrients without also ingesting toxic poisons.

FIBER AND HEALTH

One recommendation that will be familiar to most readers is to concentrate on fiber consumption—but not for the usual reasons.

The current popularity of fiber, bran, or roughage in the diet is partially based upon studies investigating why primitive peoples do not suffer from the killing "degenerative" diseases of technological societies. Most primitive peoples do eat a considerable amount of vegetables or fruit that is high in fiber, and this has been the focus of most of the researchers, beginning with Drs. Denis Burkitt and Hugh Trowell, who first suggested the link between the high-fiber diet and the absence of disease in the 1950s. But this research has paid little attention to those primitive peoples who have a high-fat, low-fiber diet yet remain just as healthy. The standard arguments for high fiber make sense but seem to leave something out.

Plant foods are formed within a fibrous cell wall composed of cellulose, hemicellulose, and lignins. None of these substances are broken down by the digestive juices, so they are not absorbed through the gastrointenstinal tract and cannot be considered essential nutrients in the human diet. The undigested plant fiber, along with a small amount of undigested protein and fat, passes through the human intestinal tract as part of fecal material. Typically, primitive Africans eat about 25 grams of fiber a day. In the average American diet, the amount of fiber consumed is very low—3 to 5 grams per day.

To explain why a high-fiber diet is beneficial, researchers compared rural African schoolchildren with English schoolchildren, whose diets offered a strong contrast by being especially low in fiber. They found that the transit time, or the time it takes for stools to pass through the intestinal tract after a meal, was considerably shorter in Africans than in English schoolchildren (for the Africans, 35 to 45 hours, for the English, 50 to 70). The researchers also found that the stools among the Africans tended to be far heavier and bulkier and were passed without effort or strain, in contrast to the "civilized" stools, which were smaller, more concentrated, and more difficult to pass. The conclusion was that the rapid passage of stools was a protective factor against intestinal diseases.

By contributing to the rapid passage of stools, fiber helps rid the intestinal tract of bacteria—especially anaerobic bacteria, which are dangerous because they can increase deoxycholic acid, a potential cause of cancer. High fiber in the diet seems to aid in preventing constipation, hemorrhoids, diverticulosis, hiatal hernia, appendicitis, colon cancers, and a host of other intestinal ills.

High fiber, then, has value as a sort of rapid-transit system (though some forms of fiber, such as gums and pectins, prolong the stool's transit time rather than speeding it). It also appears to have one other beneficial effect. Fiber acts like a glue, attracting and accumulating bile digestive juices that are the starting point for the production of cholesterol. This suggests that fiber may be important for lowering high blood levels of cholesterol.

But the flaw in the high-fiber theory is that many primitive tribes manage to avoid the "degenerative" diseases without consuming high-fiber diets. The theory cannot adequately explain why peoples such as the Eskimo and the Masai, who eat large amounts of meat and virtually no fiber or roughage, do not suffer the ailments associated with meat and cholesterol. The Chinese before 1940, and our own ancestors before the 1900s, also consumed diets high in fat and cholesterol without suffering coronary heart disease. This evidence should be an embarrassment to the high-fiber low-fat theorists, though they somehow manage to ignore it.

There is one common denominator among all these peoples, whether they are meat eaters or pure high-fiber vegetarians:

the absence of chemical additives or pollutants in their food and water supplies.

Water in primitive lands—as in developed countries before the late nineteenth century—is not disinfected. There are no industries and factories pouring waste pollutants into the immediate environment, so that plants, marine life, and land animals are not tainted by dangerous chemicals. Finally, primitive peoples have not incorporated food additives, excessive salt, bleached sugar, and bleached flour into their diet.

It has been well documented that undeveloped societies undergo a change in health patterns once they adopt the diet and lifestyle of developed societies. When the early Spaniards came to the American Southwest, they brought sheepherding to the Navajo Indians. But the change to a high-fat diet of lamb and mutton did not introduce the "degenerative" diseases to the Navajo. They continued their sheepherding for centuries after the Spaniards left, with no ill effects. In the twentieth century they acquired automobiles and other amenities of civilization. They were able to leave their Arizona and New Mexico reservations to buy food, alcohol, and tobacco products, and at that point their health patterns began to change. They now suffer from the same "degenerative" ailments as other Americans.

Clearly there are too many inconsistencies in the evidence to support high fiber as the single most important factor in prevention of the man-made diseases. When we add bran or fiber to our diet while continuing to eat all the meats and cheeses we did before, we do experience some improvement in bowel function, but not much happens to lower our blood cholesterol levels. We cannot even hope to approximate the good health of the fat-consuming Eskimos or that of our ancestors only a few generations back.

The answer lies in the chemical pollutants in our environment. But how is it possible that they can contaminate our food sources so easily?

BIOCONCENTRATORS AND DISEASE

As pointed out in Chapter 1, all animals, including humans, are bioconcentrators and biomagnifiers. Pesticides and other poisonous chemicals do enter the food chain from the soil to fruits and vegetables, but in most cases the toxins absorbed by plants remain at the same level of concentration as in the original soil. However, animals concentrate and store the poisons they eat. Many of these toxins are fat-soluble and make their home in the fatty tissue and cholesterol areas of the host animal. Biomagnification causes the amount of these poisons to increase dramatically over time, since the animal continues to ingest and store more poisons. Older and larger cattle, fish, and poultry have greater amounts of toxins than younger ones.

While the greatest concentration of toxins is in the cholesterol and fatty tissue, other tissue is also affected. Xenobiotics appear in liver and muscle tissue as well as in milk and eggs. Eating lean beef or restricting the animal food intake to fish, egg whites, nonfat milk, or skinless white meat of poultry solves little and is a compromise of dubious value.

If our meat, fish, and poultry were as free of toxins as the animal products consumed by primitive peoples in the undeveloped world, we might be able to continue to be omnivorous. However, the contamination of our soil and water now seems to be nearly total; there is little unpolluted grazing land left to us. This problem is, of course, complicated by the development of agribusiness, in which cattle and poultry are raised en masse and fattened by the use of hormones, waste tallow supplements, antibiotics, and chemically sprayed grain and feed.

Fat, cholesterol, and animal protein are not inherently bad for us; our ancestors were well able to handle them. It is the toxins we have added to them that cause the harm. For this reason, consumption of high fiber is not enough; we must also avoid foods from animal sources.

But if humans are bioconcentrators, don't we still accumulate toxic substances from fruits and vegetables? Yes, we do. But by emphasizing plant foods in our diet and eliminating animal foods, we are minimizing the dangers as much as possible. The toxin levels in animals are many times—in some studies, as much as 16 times—those in plants.

And those in plants are high enough. For example, a study by the Natural Resources Defense Council found that 44% of fruit and vegetable produce contained residues of 19 different pesticides; 42% of the samples contained more than one pesticide, some as many as four. The consumption of foods containing two or more pesticides is especially risky, for synergistic action can take place, with the toxic effects of one pesticide being enhanced by exposure to other toxins.

Pesticide monitoring is carried out daily by the U.S. Food and Drug Administration (FDA) and the Department of Agriculture. But this monitoring is simply that—monitoring. It does not prevent pesticides from being used. There are currently more than 300 pesticides registered for use on food crops. When we consider that these are used on plant food for animals and are then concentrated and magnified within the cow, pig, or chicken, interacting with the hormones and antibiotics they are given, we can begin to comprehend the magnitude of our own toxin consumption. Combine the lower level of toxins from our plant foods with the concentrated and magnified toxins from animal foods, and we are literally poisoning ourselves each time we eat.

It is true that most food contains only infinitesimal amounts of synthetic colors, preservatives, fertilizer, pesticides, hormones, and antibiotics, but most of these substances do not leave our bodies. They remain stored within us to combine with toxins from our next meal and our next, until there are enough to affect our health adversely.

The evidence is overwhelming that xenobiotics have contaminated our food and environment; they are almost certainly the culprits involved in causing coronary heart disease, cancer, Alzheimer's disease, multiple sclerosis, lupus, ulcerative colitis, rheumatoid arthritis, diabetes, and a host of other diseases.

In my own medical practice I have seen the low-toxin program help patients with such conditions as coronary artery disease, cancer, rheumatoid arthritis, gout, systemic lupus, Crohn's disease, ulcerative colitis, spastic colitis, diabetes, duodenal ulcers, chronic hives, eczema, and hypertension. My results clearly point to the urgent need for medical science to give priority to further research into the effect of xenobiotics in the genesis of these diseases.

As I've said, many of my recommendations are not unique to this program. Others have suggested variations of them to prevent or treat a variety of conditions. For example, the late Nathan Pritikin pioneered a similar program based on different reasoning. In the mid-1950s Pritikin developed progressive heart disease. Receiving little help from his physicians, he began his own research and concluded, among other things, that dietary fat and cholesterol were significant contributors to coronary heart disease. He promptly dropped them from his diet and made a full recovery. The program he originated has since helped many others.

In *Recalled by Life*, Dr. Anthony J. Sattilaro recounts his personal defeat of advanced cancer with the aid of a macrobiotic diet (a Far Eastern version of a low-toxin diet), after the failure of surgery, radiation, and chemotherapy. The National Cancer Institute and the American Heart Association also offer dietary guidelines, but these are less restrictive than the low-toxin diet, and in my opinion less effective.

For the most part, the dietary recommendations of the major medical organizations are rather tame. The medical profession has been notoriously slow in accepting the role of nutrition and xenobiotics in disease and health. Statistics show that coronary heart disease, cancer, and other conditions occur with equal frequency among physicians, their families, and the general public. It is apparent that doctors, too, need to change their lifestyle and viewpoints on nutrition.

Chapter 3

Preparing to Begin the Program

Becoming aware of toxins in our diet and environment, and taking the necessary steps to protect ourselves as much as possible from their devastating effects on our health, is perhaps the most important decision we can make to change our lives for the better, to improve the way we feel each day and the way we age.

By following the ten-point program you will secure a sense of well-being. You will begin to feel and look better than ever. If you are overweight, you will experience a gradual reduction in weight. Insomnia will diminish or disappear. Alertness will increase, and you will be better able to cope with stress. Such problems as fatigue, constipation, disagreeable body odors, and headaches will begin to clear. Other immediate effects may include heightened sexual interest and performance. Improved stamina and athletic ability are an additional dividend. The list goes on. Add to these benefits your increased chances of living a long, healthy, and happy life, and your motivation for adopting the new lifestyle this program recommends should be overwhelming.

The ten-week program is designed to introduce the ten major changes in diet and lifestyle gradually, allowing for an adjustment to one alteration before presenting the next. The easier tasks are offered first, and the most difficult hurdles (for most people) come last, after you have gained confidence in your ability to follow the program.

Throughout the program, however, remember that you have to answer only to yourself. If you have difficulty adjusting to a specific step, allow yourself more time before proceeding to the next. If you adjust quickly, you may proceed to the next step earlier, but make sure you have *fully* adjusted. If you become frustrated at making too many changes too swiftly, you may feel like giving the program up entirely.

The most important thing is to be kind and understanding toward yourself.

KICKING YOUR HABITS

Certain of the points in the low-toxin program may involve extra effort for some people. Anyone who uses tobacco, alcohol, or addictive drugs should consider these habits separately from the ten-week schedule, for breaking them involves considerable determination and may require professional help, especially if an addiction exists. Precisely when you tackle these problems is entirely up to you—but the sooner, the better.

All three habits involve xenobiotic contamination of your body. Tobacco, alcohol, and drugs have been proven to be potential causes of some of the killer diseases. Often they are a major cause.

Tobacco

Eliminating the use of tobacco products is one of the most important parts of the low-toxin program, but this is admittedly not easy to do. However, 35 million Americans are now ex-smokers, so it *is* possible for people to stop. Tobacco use is a problem not just in the United States but all over the developed world. Scientists in the Soviet Union consider it of such importance in their country that they have called for the abolition of all tobacco products by the year 2000.

Smoking tobacco causes xenobiotic damage in two ways. In its gaseous phase, when the smoke is puffed and exhaled, it is of risk both to the smoker and to those around him who inhale secondhand or sidestream toxins. Tobacco smoke contains benzo[a]pyrenes, formaldehyde, nitrosamines, hydrogen cyanide, aromatic hydrocarbons, phenols, and polonium 210, a radioactive element. All of these are known carcinogens and xenobiotics.

In addition, smoke contains carbon monoxide, which can contribute to coronary heart disease and stroke, and nicotine, a stimulant that can raise blood pressure, speed the heart rate and pulse, and increase the speed of transmission of nerve impulses. Cigarettes also contain additives that are considered trade secrets and are not disclosed to the government or to the public, and these may involve additional toxins.

The second kind of xenobiotic damage from smoking tobacco occurs in the tar phase. Here the damage is slower and more prolonged, resulting from tar accumulating in the lungs for years, silently causing free-radical damage while contributing to the development of cancer, bronchitis, or emphysema.

Because of the accumulation of tar, two-pack-a-day smokers double their chances of suffering heart attack and increase their chances of developing lung cancer twentyfold. Statistics show that 30% of all cancers and 82% of lung cancers can be attributed to the xenobiotic action of tobacco smoke. People who have smoked a pack to a pack and a half a day for 10 years or longer will need at least 15 years of nonsmoking before their risk of cancer is equal to that of nonsmokers, and 10 years to similarly reduce the risk of coronary heart disease. However, recent reports indicate that the risk of a heart attack decreases almost immediately after the cessation of smoking.

The fact that giving up smoking decreases the risk of cancer and heart attacks is encouraging. It suggests that enzymes can detoxify xenobiotics slowly over time and the body can repair much of the damage. At some point, however, the damage becomes irreversible. It is vital to make the necessary changes before that point is reached.

It is important to remember that smoking affects not only the smoker but those around him as well. Nonsmoking wives of cigarette-smoking men have been shown to live an average of four years less than wives of nonsmoking husbands. People who are subjected to sidestream smoke—that which results from letting a cigarette burn slowly, unsmoked, as in an ashtray—are more likely to suffer from episodes of angina or chest pain.

The effects of secondhand tobacco smoke are particularly hazardous for children. The children of smoking parents (especially of smoking mothers) are more likely to develop respiratory illnesses such as bronchitis, pneumonia, and asthma than are the children of nonsmoking parents. Pregnant women who smoke are endangering the lives of their unborn infants. Not only do smoking mothers tend to have smaller babies than nonsmoking mothers, they also have a higher incidence of infant death and other complications at the time of delivery. Women should stop smoking before becoming pregnant or, at the very least, stop as soon as they learn they are pregnant. Ideally, any woman who smokes and is in her childbearing years should stop immediately.

But if we are speaking of ideals, no one should smoke in the first place. If you do not smoke, don't start; if you do smoke, you should do anything possible to stop.

There are strong indications that the toxins in cigarette smoke have a magnifying effect on other toxins we get from tap water, food, and exposure to air pollution, radiation, chemicals, and dust. Since you are reading this book, you obviously care about life; but if you smoke, it makes little sense to undertake the low-toxin program without making an effort to quit.

And it is not just cigarettes, cigars, and pipe tobacco that should be avoided. Chewing tobacco and snuff are equally dangerous, though the dangers have not been as highly publicized because these products have been less popular. Recently, however, chewing tobacco has come into greater use among professional athletes, and as a result the habit has spread to the youngsters who look up to them. Consequently, there has been an increase in the incidence of mouth cancers, which are caused by the nitrosamines in tobacco.

Tobacco is addictive, and it is extremely difficult to give up an addiction. Certainly it is not as easy to stop as it is to start. Most smokers begin in their teen years, when peer pressure is great. For those who have smoking parents, peer pressure is reinforced by their role models. Added to this, thanks to advertising, movies, and television, is the perceived glamour of smoking. Once acquired, the habit becomes ingrained and automatic in adulthood. Drink a cup of coffee for breakfast, and the smoking begins for the day. Sit down at the desk to make phone calls, and a cigarette seems necessary. After work, during happy hour, smoking goes hand in hand with the drinking so essential to "sociability."

The experts have been unable to agree whether smoking is a physical addiction, a psychological dependency, or some combination of both. This dispute may not be important; what is truly important is that the addiction be overcome.

Whatever its basis, smoking is a difficult addiction to conquer. One technique for quitting may work for one person but not for someone else. Though specific recommendations are beyond the scope of this book, there are many programs available. Select one that feels right for you, and if it doesn't work, try another. It's worth the effort.

Both the American Heart Association and the American Lung Association have booklets, brochures, and programs dealing with smoking. Many community hospitals and churches sponsor seminars on the subject. You may wish to consult your physician for assistance; one thing he or she can do is to prescribe a nicotine gum that is intended to satisfy your nicotine cravings and is easier to give up than cigarettes. Hypnosis and self-hypnosis techniques, as well as group therapy, individual therapy, and aversion techniques, are available through psychologists, counselors, and psychiatrists. Commercial behavior modification programs such as the Schick Centers and Smokers Be Free advertise high success rates, and there are indications that they do have some success. "Cold turkey" works for some people, and others claim positive results from acupuncture and pins in the earlobes. See if your place of work has a counseling program.

Important to any program you undertake is that you must truly want to quit smoking. Have the determination to succeed, and you will. Keep trying as often as necessary until you *do* succeed. It is truly amazing how much healthier and more vibrant you will feel once you lick this habit. Ask any ex-smoker.

Alcohol

Doctors disagree about the value of alcohol consumed in limited amounts. Some reports indicate that alcohol in small amounts raises levels of high density lipoproteins (HDL), which are presumed to protect against coronary artery disease. However, it raises levels of HDL3, a subgroup of high density lipoproteins that do not offer this protection (it is HDL2 that decreases the risk of coronary disease).

Any potential beneficial effects of alcohol are far outweighed by its harmful ones. Along with tobacco, alcohol ranks as one of the most significant contributors to health problems. The aldehydes, ketones, fusel oils, polymers, and other solvents in alcohol cause liver damage, cancer, birth defects, elevated triglyceride levels, high blood pressure, and stroke.

But the most serious concern with alcohol is that it is addictive. Terrible damage results from excessive drinking among both adults and teenagers. The risk of heart disease, cancer, and cirrhosis of the liver increases dramatically with increased alcohol consumption. Just as with tobacco, alcohol overwhelms vitamins. Not only do levels of vitamin C decrease, but vitamin A, niacin, and thiamine deficiencies may develop, and these can cause conditions such as pellagra, beriberi, heart disease, and damage to the nervous system.

Anyone who is addicted to alcohol or who uses it in excessive amounts may need to seek professional help as part of the low-toxin program. The amount of alcohol that is considered excessive varies, depending upon a person's weight and how often he or she drinks. To an alcoholic, even a few drops is too much. Certainly, someone who imbibes alcohol infrequently or has an occasional glass of wine or a beer or cocktail is not a problem drinker. If you are not a problem drinker and you stick to the low-toxin program, you can reward yourself with one or two drinks at important social occasions, if you desire.

For the problem drinker there are self-help and professional treatments available, including Alcoholics Anonymous, Schick Centers, the Betty Ford Center, and hospital programs.

Addictive and Other Abused Drugs

The major drugs used today include marijuana, the hallucinogens LSD and mescaline, PCP, cocaine, sedatives, and narcotics. All have xenobiotic potential and can cause damage to the human body.

Marijuana is derived from a common plant called *Cannabis sativa*. Its chief active ingredient is delta-9-tetrahydrocannabinol (THC). Marijuana is far more dangerous than was suspected only a few years ago. The use of "pot" causes symptoms such as increased pulse rate, red eyes, and dry mouth and throat. Those who use it seek a feeling of euphoria and relaxation, but it may also result in impaired memory, an altered sense of time, and reduced ability to concentrate. Coordination, reaction times, and ability to cope with stress are badly affected.

There are long-range dangers too. The smoke from one "reefer" contains about the same amount of cancer-causing chemicals as the smoke from one pack of cigarettes. People with coronary heart disease are at risk for developing chest pain or heart attacks, since the pulse rate may accelerate by up to 50% and the blood supply to the coronary arteries will be diminished.

The psychedelic drugs LSD (lysergic acid diethylamide), which comes from the fungus ergot, and mescaline, derived from the peyote cactus, cause altered vision, hallucinations, and illusions. Even after use of these drugs is discontinued, flashbacks can occur, and there is danger of long-term anxiety, depression, and permanent brain damage.

The feeling of euphoria is also what leads people to experiment with PCP (phencyclidine), also known as "angel dust." This insidious drug can cause muscle rigidity, loss of concentration and memory, convulsions, and violent, bizarre behavior.

One of the most popular and vicious drugs today is cocaine, a stimulant extracted from the leaves of the coca plant. Formerly, its primary use in medicine was as a local anaesthetic, particularly in nasal surgery. Sigmund Freud administered it to patients in the last century in the misguided belief that it could help to eliminate morphine addiction.

Street cocaine is usually sold as a white crystalline powder or in small pieces. "Coke" is commonly inhaled, or "snorted"; some users swallow it, inject it, or smoke a highly purified form of the drug called freebase. Freebasing is a less long-lasting but more intense way of using the drug. And if the effects are more intense, so are the dangers. In recent years, the use of "crack" or "rocks," a derivative made by heating cocaine, baking soda, and water, has increased the risks of almost immediate addiction.

The dangers of cocaine include high blood pressure, heart attacks, convulsions, psychosis, hallucinations, and mucous membrane and nasal septum destruction, even after cessation of the habit. It is suspected that cancer of the liver can occur after long-term use, since cocaine is frequently contaminated with benzene, a known carcinogen.

Stimulants such as amphetamines have been used in medicine to help suppress appetite, though they have not been very effective as aids to weight loss. They have also been used to increase alertness and sharpen physical ability. But amphetamines have their dangers too. They can cause restlessness, insomnia, mood swings, high blood pressure, acne, and brain damage. They can also induce heart attacks in high-risk people.

The sedatives include prescription tranquilizers and sleeping pills, sometimes nicknamed "downers," "reds," or "barbs." The most commonly abused barbiturates are pentobarbital

(Nembutal), secobarbital (Seconal), and amobarbital (Amytal). These are all used medically as sleeping aids. The nonbarbiturate sedatives include glutethimide (Doriden), meprobamate (Miltown), ethchlorvynol (Placidyl), and diazepam (Valium).

Sedatives are used medically to induce sleep or relieve anxiety. However, it must be noted that they are addictive and dangerous. The long-range effects are no different than those of other xenobiotics. In the short term they can kill. Barbiturates are involved in a third of all drug-induced deaths. In combination with alcohol, they are deadly.

The narcotics opium, heroin, morphine, and meperidine (Demerol) are pain relievers, and some of them have a place in medicine. But these drugs are also addictive and dangerous. Moreover, the use of contaminated drugs, syringes, and needles can lead to major disorders such as hepatitis and AIDS, sometimes through the transmission of viruses but also because the drugs may contribute to partial suppression of the immune system. Life expectancy among narcotics users is low.

CONSULTING YOUR DOCTOR

Before beginning this health regimen—or any other—it is advisable to pay a visit to your doctor for a physical examination. Even if you feel perfectly healthy, you should determine whether there are any early warning signs of disease. If there are none, starting and maintaining the program should keep you in good health. You should discuss your plans to go on this program with your doctor. It is important for him or her to be familiar with the concepts of this book, or to be willing to learn about them, so that you can derive the maximum benefit from the examination and the ten-point program. If necessary, loan this book to your doctor. After all, the ideas it presents are so new that physicians are unlikely to have heard them in medical school. (The bibliography of research papers at the back of the book may be of particular interest to doctors.)

There was a time when most patients faced their doctors in awe of their medical powers over life and death. But times have changed. Now we are likely to be judged as human beings with specialized knowledge not unlike that of lawyers, accountants, and other professionals—and we are held just as accountable. Patients have a perfect right to engage us in dialogue, especially in matters as vital to them as their health—and of course they also have the right to get a second opinion from a doctor who may be more in sympathy with the ideas in this book.

More and more doctors are becoming aware of the health problems posed by the nutritional hazards of our modern world, but admittedly the number is still small, and you may have trouble finding such a doctor in your area. The Directory of Selected Resources at the back of the book lists organizations and publications that may help you find a physician who meets your needs. However, you should be aware that the ideas in this book are very new; it may take a while before they are recognized and accepted by physicians—even those attuned to the relationship between nutrition and disease. I have also listed my address in the event you wish to contact me for help in finding a physician in your area, or to obtain my newsletter or ask questions about the ten-point program.

Health History

Your checkup should include a history of your health, a physical examination, and laboratory tests. The health history involves a listing of all illnesses and physical complaints you have experienced throughout your life. You will be asked to answer questions concerning the health, diseases, and causes of death of relatives, for some disease conditions tend to run in families. (Of course, with the low-toxin program, many of them can be prevented.)

Physical Examination

You should be examined thoroughly. Women should have breast and pelvic examinations to check for early signs of cancer. (In addition, women should learn how to examine their own breasts and do so on a monthly basis, in between their annual doctor's checkups; but they should also be aware that recent studies have cast doubt on the effectiveness of breast self-examination and that mammograms may prove to be more useful.) Inescapably, cancer is a major concern in developed societies, and everyone should be examined and tested on a regular basis (see Table 3.1).

Regular physical exams are very important, but it is most important to avoid cancer by all means possible, including the low-toxin program. Do not rely on the checkup to detect cancer early enough to be truly helpful. There are more than a billion cancer cells in a barely detectable tumor the size of a thumbnail. It may have taken an average of ten years to reach this stage of growth, and over that period cancer cells may have colonized elsewhere in the body. Early detection examinations will not find these colonies. *Prevention of cancer must begin long before the checkup.*

Laboratory Tests

Whatever your age or sex, you should have certain routine laboratory tests. For adults over the age of 40 and for those in high-risk categories, additional tests are necessary. The standard tests should be performed even on apparently healthy teenagers. In my routine examination of several hundred high school athletes, high blood pressure was discovered in a surprising number who seemed completely healthy and had shown no previous signs of the problem. Blood cholesterol levels may also be rising in teenagers.

A urinalysis and standard blood chemistry screening are very important tests for everyone except infants and young children. The tests should measure blood sugar, uric acid, kidney and liver function, blood salts (electrolytes), iron, cholesterol, triglycerides, and high-density lipoprotein, as well as hemoglobin, white blood cell count, and red blood cell count. Total blood cholesterol for an average adult should be below 180 mg/dl; triglycerides should fall below 125 mg/dl. Ask your physician to explain your test results to you, and be sure to get a copy for future comparison.

If there are any abnormal findings, particularly in uric acid, cholesterol, glucose, or triglyceride levels, repeat examinations should be performed every three months until values return to normal. Of course, if you do not follow the dietary recommendations of the ten-point program, your blood test results will probably remain unchanged.

Adults over the age of 40 should have an electrocardiogram, a test to look for hidden blood in the stool, and possibly a sigmoidoscopy to detect possible colon cancer. (But again, do not rely on this form of "early detection" as an effective means of preventing cancer. For meaningful cancer prevention, you must follow the low-toxin program.) Treadmill stress tests should not be routine but should be performed for people in whom heart disease is suspected.

One test that should never be done routinely is a chest X-ray. For the average person, the risks far outweigh the advantages, though X-rays may be necessary for high-risk people like smokers, recovering cancer or tuberculosis patients, and those with certain other ailments. In specific cases your physician may require upper and lower GI series or kidney X-rays for diagnosis or medical decisions; here the risk of the X-ray is outweighed by the potential seriousness of the ailment.

Finally, we should all get into the habit of talking to our physicians and learning their attitudes toward disease prevention and the prescribing of drugs. What is your doctor's attitude toward nutrition, antioxidants, and exercise? Is your doctor a smoker? Overweight? Is your doctor easily approachable and able to answer all your questions satisfactorily?

Questions to Ask Your Physician
- Is the exercise program in this book (Chapter 6) acceptable for me? Do you have any other suggestions for an exercise program?
- Do I have any disease or ailment that will be complicated by taking the vitamin supplements recommended in Table

TABLE 3.1

Summary of American Cancer Society Recommendations for the Early Detection of Cancer in Asymptomatic People

Test or Procedure	Sex	Population Age	Population Frequency
Sigmoidoscopy	M & F	Over 50	After 2 negative exams 1 year apart, perform every 3–5 years.
Stool guaiac slide test	M & F	Over 50	Every year
Digital rectal examination	M & F	Over 40	Every year
Pap test	F	20–65; under 20, if sexually active	After 2 negative exams 1 year apart, perform at least every 3 years.
Pelvic examination	F	20–40	Every 3 years
		Over 40	Every year
Endometrial tissue sample	F	At menopause, women at high risk[a]	At menopause
Breast self-examination	F	20 and over	Every month
Breast physical examination	F	20–40	Every 3 years
		Over 40	Every year
Mammography	F	35–39	Baseline
		40–49	Every 1–2 years
		50 and over	Every year
Chest X-ray			Not recommended
Sputum cytology			Not recommended
Health counseling and cancer checkup[b]	M & F	Over 20	Every 3 years
	M & F	Over 40	Every year

Note: Reprinted by permission of the American Cancer Society from *Ca—A Cancer Journal for Clinicians* 35 (July–August 1985): 199. © American Cancer Society.

[a]History of infertility, obesity, failure to ovulate, abnormal uterine bleeding, or estrogen therapy.

[b]To include examination for cancers of the thyroid, testicles, prostate, ovaries, lymph nodes, oral region, and skin.

5.1? If I do, could you provide me with the references so that I can read the research material?

- May I try the low-toxin diet, nicotinic acid, vitamin E, and omega-3 marine lipids to lower my cholesterol level, instead of the drug I am now taking?
- Is it possible to stop all my medications gradually if I improve with this program? If my blood pressure becomes normal? If my diabetes improves? If my cholesterol drops below 180 mg/dl? If my arthritis clears?

To my knowledge, there are no health restrictions on starting the ten-point program. It is a program for all ages and for people in various states of health, from the chronically ill to the healthy, well-trained athlete. Everyone can benefit, whether they intend to remain novices and follow a few guidelines selectively or to go all out and adhere strictly to the precepts.

Part II
The Program

THE TEN-POINT LOW-TOXIN PROGRAM

1. Avoid tobacco, and restrict or eliminate the use of alcohol.
2. Drink pure water, preferably distilled water.
3. Take antioxidant vitamin and mineral supplements.
4. Develop and follow a regimen of exercise.
5. Keep your weight down, since body fat is a storage area for toxins.
6. Avoid processed foods.
7. Avoid animal food sources.
8. Base your diet on plant sources for food.
9. When possible, avoid drugs of all types (illicit, prescription, and nonprescription), radioactive isotopes, and X-rays.
10. Be aware of the toxic dangers in your environment and minimize those you can control.

Week One: Water and Beverages

Projects:
1. Prepare a shopping list.
2. Obtain distilled water or a water distillation system.
3. Start drinking healthier beverages.

The changes you will be making in the first week of your program are relatively easy, yet they are among the most important. Do not expect any of the program to be effortless. You will have to do some searching through your Yellow Pages this week and make some extra purchases at your supermarket. You will also be examining your kitchen to begin to weed out some products and remove them from your shopping list. You are going to be taking a good look at your lifestyle, discovering some of the dangers that surround you, and getting rid of them.

YOUR SHOPPING LIST

A good first step for your first week is to make a shopping list, not one based upon my recommendations, but one consisting of all the products you normally buy at the supermarket. If some other member of the household usually does the shopping, you will need to enlist his or her aid. Include everything, but separate foods from household products. Do a good job— type it up if you can—for you are going to be using this list throughout the entire ten weeks of the program, removing some items and adding others. Keep it handy; your refrigerator door is a good place.

There are several approaches to making your list. You can mentally take yourself through your supermarket, recalling what you put in your basket from each section. Or you can look through your cupboards and refrigerator to see what products you presently have on hand and what you think you currently "need." You can also refer to Table 4.1, which may help to prod your memory.

As this week progresses, you will be removing several items from your shopping list. You may also be adding some things. The major alterations at first will involve beverages, milk products, and eggs. If your supermarket sells the right kind of water, you may choose to add it to your shopping list as well. The emphasis this week is on water and beverages.

WATER

Water is an amazing substance. Although it does not supply energy for our bodies and does not go through a digestive process before being absorbed, water is an essential nutrient. Along with oxygen and energy from the sun, it is one of the three prerequisites of most forms of life. Human beings cannot exist without adequate water.

Water comprises a very large percentage of our body weight. At birth, a baby's body is approximately 77% water. The percentage gradually decreases with age, but for adult men and women it is still between 45% and 65%. Roughly, this comes to about 12 gallons of water per person, and to maintain this amount over our lifetimes we have to consume approximately 10,600 gallons.

Thus, the purity of the water we drink should be extremely

TABLE 4.1
Typical American Shopping List

Foods		
Produce Section		
lettuce	broccoli	melons
tomatoes	asparagus	corn
potatoes	oranges	garlic
onions	tangerines	chives
green onions	limes	carrots
radishes	lemons	beets
beans	grapefruit	cabbage
peas	nectarines	brussels sprouts
cucumbers	peaches	mushrooms
bean sprouts	plums	endive
artichokes	apricots	sweet potatoes
parsley	apples	cauliflower
peppers	pears	squash
greens	pineapples	zucchini
avocados	kiwi fruit	spinach
celery	bananas	
Dairy Section		
eggs	cream cheese	sour cream
homogenized	cheese spread	whipping cream
milk	butter	whipped cream
low-fat milk	margarine	imitation whipped
nonfat milk,	lard	cream
raw milk,	fruit juices	half-and-half
kefir	party dips	nondairy creamer
Jalisco cheese	creamed or pickled	low-cholesterol
Brie	herring	"artificial eggs"
Camembert	salted herring	yogurt
Gorgonzola	smoked fish	nonfat yogurt
cottage cheese	luncheon meats	flavored yogurt
farmer's cheese	lox	buttermilk
hoop cheese		

(continued)

TABLE 4.1 (Continued)
Typical American Shopping List

Foods

Bread Section

white bread	corn tortillas	raisin bread
pumpernickel	rye	English muffins
corn bread	whole wheat	hamburger and hot-
sprouted-grain	sourdough	dog buns
bread	French rolls,	donuts
pita	dinner rolls	pretzels
whole wheat pita	Kaiser rolls	cakes
flour tortillas	bagels	cupcakes

Spices, Condiments, and Baking Goods

spices	cooking oil	whole wheat flour
salt	olive oil	cornstarch
sugar	shortening	cornmeal
monosodium	vinegar	cake and cookie
glutamate	powdered chocolate	mixes
bottled salad	presweetened drink	bread mixes
dressings	mixes	pie fillings
dehydrated soups	bleached white	gelatin
mayonnaise	flour	pasta
mustard	unbleached flour	noodles
ketchup		

Canned and Bottled Goods

vegetables	potted meat	tomato puree
fruits	products	tomato paste
vegetable juices	olives	spaghetti sauces
fruit juices	pickles	menudo
fish	pastas	chili
soups	tomato sauce	chili peppers
beef stews		

TABLE 4.1 (Continued)
Typical American Shopping List

Foods

Frozen Foods

ice cream	premixed breads	hash browns
sherbet	and cookies	berries
frozen dinners	fruit juice	pies
"dietetic" dinners	concentrates	frozen desserts
pizza	french fries	cakes
vegetables		

Party Supplies

soft drinks	dip mixes	spice tea
flavored drinks	candy	wine
chips	chocolates	liquor
crackers	coffee	beer
cookies	tea	

Cereals

corn flakes	rice cereals	cornmeal mush
bran flakes	sugared cereals	grits
shredded wheat	hot wheat cereals	oatmeal

Meat Section

beef	turkey	sweetbreads
pork	turkey roll	brains
lamb	turkey ham	heart
chicken	turkey pastrami	kidney
barbecued	ham	lungs
chicken	pastrami	fish
barbecued spare	corned beef	bacon
ribs	calf liver	frankfurters
smoked pork	beef liver	breakfast sausage
ham hocks	pig liver	English bangers
chicken livers	pâté	Italian sausage
duck	tripe	bratwurst
squab	pig feet	chorizo
Cornish hen	pig knuckles	Portuguese sausage

(continued)

TABLE 4.1 (Continued)
Typical American Shopping List

Household Products

paper products	water softener	ammonia
powdered cleanser	fabric softener	bath soap
liquid cleanser	drain cleaner	tile cleaner
laundry detergent	chlorine and	oven cleaner
soap	nonchlorine	ant spray
spot remover	bleach	roach powder
washing soda	oven cleaner	mouse traps
dishwasher	furniture polish	rat poison
detergent	floor wax	

important to us. Water can carry toxic chemicals in suspension along with valuable minerals. And we have been consuming numerous man-made chemicals at least since the early nineteenth century. After the cause-and-effect relationship between various illnesses and bacteria was discovered by Louis Pasteur and Robert Koch, new chemicals were introduced to "purify" water supplies. These disinfectants are another form of contamination contributing to the development of new, potentially more serious diseases.

Public Water Supplies

Between the layers of soil on the earth's surface and the deep bedrock are layers of clay and porous limestone or sandstone that form the foundation of aquifers—the underground storage areas for much of our usable water. At least 50% of the earth's population uses the water from aquifers, called groundwater, rather than that from lakes, rivers, and streams. This groundwater exists in billions of acre-feet of underground streams and lakes.

We make use of groundwater by tapping natural springs or by drilling wells and pumping it to the surface. There appears to be a never-ending supply of this water, but actually, as we withdraw it and use it up, it is only partially replenished by rain or melting snow from the earth's surface.

The rainwater that reaches the aquifers may contain unwanted substances—the chemicals in acid rain, produced by industrial smoke; radioactive fallout from nuclear tests or accidents; and lead from automobile exhausts. Aquifers also become polluted through other processes. Agricultural runoff from irrigation seeps into the soil loaded with nitrates from fertilizer, both natural and synthetic, which percolate and migrate to the aquifers, as do chemical pesticides, rodenticides, and insecticides. Some of these pollutants even reach the sea, ending up as part of the ocean environment.

Toxic dump sites like Love Canal, Times Beach, or the Stringfellow acid pits have contributed to the underground plumes of moving toxins that eventually reach the aquifers. Even individual septic tanks and garbage from sanitary landfills contribute leakage, as do the literally millions of new chemicals devised by industry; over 66,000 different compounds are prepared each year, of which less than 2% have been checked for possible side effects.

From time to time pollution accidents such as those at the Three Mile Island and Chernobyl nuclear plants, the catastrophic chemical spill into the Rhine River, and the escape of methylisocyanate in Bhopal, India, occupy the attention of the news media. They are newsworthy because the results of the disasters can be seen immediately, affecting people in the surrounding countryside. However, the larger disaster is not so immediate; it occurs quietly after years of rainwater seepage of chemicals and radiation in combination with seepage from the burial of PCBs (polychlorinated biphenyls), dioxin, and radioactive wastes. The numerous cancers, birth defects, and other man-made diseases that show up years later, often miles from centers of disasters, will not hit the news.

Chemicals and radioactive substances remain in our water supply. Public water systems then add chlorine to make it germ-free. They may also add up to 50 other chemicals, but not all of these are used in all water plants at all times; some are used only at specific times for specific purposes—to adjust acidity or alkalinity, for algae control, for corrosion and scale control, for water softening, for taste and odor control, for disinfection, and so on. Water can pick up additional pollutants on its way to your home. Water is a solvent, and as it travels through plastic, asbestos-lined concrete, or metal pipes containing soldered joints, it can absorb polyvinyl chloride, asbestos, lead, and cadmium, all of which are toxic. This is a problem even in those areas of the country that pride themselves on "good" water.

One family of chemical additives to our water has become a very sensitive issue. Even to suggest that there are hazards with fluorides is to risk being categorized as a member of the lunatic fringe. However, the subject must be considered rationally and seriously.

Widely found in nature, fluorine is the thirteenth most abundant element in the earth's crust. It is a highly reactive substance, with a strong tendency to combine with other elements to produce fluorides. It is a halogen, part of the same family of elements as chlorine, bromine, and iodine, which are among the most highly reactive poisons known.

Since fluorides exist throughout the environment, they can enter the body from the air, foods, and water, even when they are not deliberately added. Many mountain spring waters contain a natural and substantial supply of fluorides.

After the discovery early in the twentieth century that fluorides could help prevent tooth decay, it was suggested that they be added to drinking water. Fluoridation trials began in 1945 in a number of cities, and after positive results fluorides began to be added to water supplies throughout the world, in amounts considered harmless. Today they are also added to foods as preservatives.

But the widespread use of fluorides and fluorine compounds for a variety of purposes does not change the fact that they are toxic substances. Some are even used as insecticides and rodenticides because of their extreme toxicity. It has been proven that excess fluorides can damage the genetic material of both plants and animals. The incidence of Down's syndrome, which is caused by chromosome disruption, is much higher in cities that have water fluoridation than elsewhere. Although fluoridation is only one factor involved, its significance should not be dismissed.

More than 300 cases of acute fluoride poisoning have been recorded, but it is likely that the reported number of cases would be much higher if it were not so difficult to diagnose the condition. Excess fluorides can cause skeletal fluorosis, with marked increase in bone density and fragility. Additional symptoms are loss of appetite, nausea, vomiting, diarrhea, constipation, aches and pains, fatigue, headache, dizziness, rashes, shortness of breath, palpitations, and cough. Kidney disease and death have occurred in some cases.

There is also evidence to suggest that fluorides are carcinogenic or at least encourage the development of cancer. Since fluorine is one of the most poisonous substances known—even more so than chlorine—this is not surprising.

The amounts of fluorides added to most water supply systems do not approach toxic levels. And except in extremely hot climates where water consumption is very high, it is unlikely that people would drink enough tap water to endanger their lives. However, in addition to fluoride from drinking water, we get fluorides and fluorine compounds from other sources, at levels that are usually not measured, with effects that are impossible to predict. Since fluorides exist throughout nature and have proliferated through industrial use in superphosphate fertilizers, plastics, refrigerants, toothpaste, and medicinal products, it is quite possible that people are exceeding safe levels of intake.

Fluoride helps to prevent dental caries only during the childhood years and offers little benefit to adults, so it seems foolish to continue adding it to drinking water. However, it is unlikely that fluorides will be removed from public water supplies. The absence of fluorides is one of the benefits of drinking distilled water, particularly for adults.

Sources of Pure Water

Obtaining pure water is not as difficult as it might at first appear. Certainly it is not as easy as turning on the tap in your kitchen, but pure water is available.

What you want is distilled water or pure artesian or spring water. If you obtain it from commercial sources, try to get it in glass containers, because chemicals in plastic will leach into the water. Unfortunately, less than 2% of bottlers today will provide glass bottles because of weight, cost, and breakage. But bottled water even in plastic is preferable to tap water as you begin the program. The best solution, as suggested later, is a water distillation system for your home. To get a good start on the program, however, you will find it easier to purchase water at your supermarket or subscribe to a water delivery system. This is important not only at home but also at work. If you own your own business, arrange for delivery of distilled water; if you are an employee, try to persuade your employer to make the change.

BOTTLED WATER

There has been a phenomenal increase in the sale of bottled water in the last 10 to 15 years. Much of this increase can be attributed to greater health awareness, and some to problems with the taste of tap water. In cases of "prestige" waters, there is also some snob appeal.

It is important to realize that not all bottled waters are an improvement over tap water. For example, some of the mineral waters contain arsenates and lead. In one test, Appollinaris Water from Bad Neuenahr-Ahrweiler, West Germany, was found to exceed the Environmental Protection Agency (EPA) tap water standard for arsenic by 6,000%, with excessive amounts of selenium and cobalt as well.

The problem with bottled water is that it is difficult to distinguish the good bottlers from the bad. An estimated 324 bottling plants in the United States produce over 400 different brands of water, and there are 35 imported mineral waters.

Because bottlers are essentially self-regulated, it is impossible to make an overall assessment of the entire industry.

In 1972, the U.S. Food and Drug Administration (FDA) conducted a survey of 25 bottling establishments that produced approximately 50 bottled water products. Tests for bacteriologic and chemical content indicated some very serious problems. Coliform organisms were found in 8% of the samples examined. During 63 days of storage, further testing revealed increases in bacteria. The study discovered that quality control measures were generally deficient and that chemical analyses of the water were not regularly performed. In many cases, bottling was not done under sanitary conditions. Plastic bottles arrived at the plants in cardboard cartons, without caps, so that their interiors were exposed to airborne contamination; yet these bottles were not even rinsed before being filled with the product water. Also, discrepancies were found between the products' actual chemical composition and that stated or implied on the labels. Chemical analysis revealed traces of chlorine, sulfate, nitrate, copper, manganese, lead, zinc, mercury, and arsenic in virtually all the samples, even those that had been subjected to filtering, distillation, and other methods of purification.

A study conducted in 1980 by an independent group indicated that there had not been a significant improvement since the FDA's 1972 study. This group tested 14 kinds of bottled water, both carbonated and noncarbonated. Some of the waters tested were of very good quality, but others contained high levels of sulfate, calcium, magnesium, and other constituents.

Be careful not to be misled by the salespeople at bottled water companies when you set about making your selection. You do *not* want fluoridated water. The first choice is *not* mineral or mineralized water, natural water, naturally sparkling water, purified water, or spring water, and you do *not* want plastic containers if glass is available. What you *do* want is distilled water, and if that is not available, you may accept artesian or spring water. It is possible to purchase distilled water bottled in glass in many health food stores. Or you can consult your Yellow Pages and contact the bottlers in your area to see if any will provide distilled water in glass.

HOME WATER PURIFICATION

In general, purchasing bottled water is the easiest way to acquire pure water as you start this program. When you have adapted to this change, you may ultimately decide you want a home water purification system. To help you select one, the following pages give a general explanation and assessment of the kinds of systems or processes that can be utilized at home. My experience suggests that distillation is the best method of water purification, but you may prefer a different method.

Boiling

Boiling is no longer a safe way of purifying water for home use. While it can be effective in ridding water of some bacteria, it is ineffective for many viruses, and it can actually concentrate nonvolatile chemicals that have been added to the water supply. Boiled tap water should *never* be used for infant formulas or drinking water. Boiling is not recommended except when unavoidable, as when camping out.

Charcoal Filtration

The most common water filters sold for home use are charcoal filters, and they are usually relatively inexpensive. However, as purifiers, carbon filters leave a great deal to be desired. They may be successful in trapping undissolved solids, some bacteria, spores, and other insoluble impurities, and they can clear murky water and remove objectionable odors. Some toxic substances, such as organic pesticides, chloramine, and chlorine, may also be temporarily trapped by carbon. However, charcoal filters do not get rid of such undesirable dissolved trace minerals as salts of arsenic, cadmium, lead, strontium, nitrates, and mercury.

In order for carbon filters to be effective in removing compounds such as pesticides and chloramines, the carbon must have adequate adsorbing power and must be fresh, closely packed, and of adequate size. Small faucet attachments of carbon are worthless. Water must come into contact with the filter for an extended time—at least three minutes, preferably longer.

Carbon filtration can be worse than no purification system at all. After a short while the carbon can become saturated with trapped material without your realizing it. For example, a very serious problem arose when chloramine was introduced as the new water disinfectant recently in California; kidney dialysis patients were adversely affected because the chloraminated water easily passed through carbon filters.

When bacteria are trapped by these filters, they can multiply in the wet carbon, with nutrients in the filter serving as a culture medium to stimulate their growth. Eventually the filter cannot hold them back, and they will contaminate the water passing through. Although most bacterial spillover contamination from carbon filters is harmless, the bacteria that cause Legionnaire's disease can survive, as can the hepatitis and polio viruses.

Charcoal filtration is not recommended for home use. It may be adequate in hospitals or industry if checked daily by technicians trained to recognize a filter failure.

Water Softeners and Conditioners

In some areas, water contains calcium and magnesium ions that make it "hard." To obtain water that is better for laundering and washing, the calcium and magnesium ions are removed through a process known as ion exchange. This "softened" water lathers more easily with soap.

However, softened water is not recommended for people with heart, kidney, or hypertension problems, because of the loss of calcium and magnesium and the addition of sodium. It is important to have an alternate source of water for drinking and cooking if you use a water softener.

Deionization

Deionization is a procedure similar to water softening, but with the addition of an anion exchange resin to remove all the salts. It is used in industry to provide high-grade rinse water with low conductivity. However, deionization does not remove organic compounds or bacteria, and it is neither practical nor useful for obtaining home drinking and cooking water. Many bottled water companies use deionization, usually in conjunction with other methods of purification such as ozonization or ultraviolet irradiation.

Reverse Osmosis

Reverse osmosis is a filtration process utilizing a semipermeable membrane. There are literally dozens of chemical polymers that can be selected for use in the membranes, each with specific advantages and disadvantages. The membranes may vary widely in their ability to reject chemicals. Theoretically reverse osmosis can reduce the total dissolved salts in water by 90% to 95%, and the rejection of viruses, bacteria, and pyrogens (fever producers) can approach 98%. However, much of the success of the method depends upon the selection of a proper membrane, which requires knowledge and judgment.

Hospitals and industries generally have technical staff with the training to use reverse osmosis effectively. Even so, it is almost never used without some ancillary method of filtration or water treatment. Though some reverse osmosis systems are sold for home use, they are usually ineffective without a technician to service them on a daily basis. In addition, they require high pressure and pretreatment of feed water—dechlorination, adjustment of acidity, and use of carbon prefilters, softeners, and chemical scale inhibitors—and it may take 7 to 15 gallons of feed water to produce 1 gallon of usable water; the rest goes down the drain. Finally, the storage tanks of home systems are made of plastic or fiber glass, neither of which provide a proper environment for water purity; glass or stainless steel is preferable.

For all these reasons, reverse osmosis is not recommended for home use.

Ozonization

Ozonization appears to hold much promise as a purifying process, though it also has some shortcomings. Most people are aware of ozone as one of the components of smog and as a protective layer of the stratosphere. Ozone is a highly oxidative free radical, and that is what makes it both valuable as a purifier and potentially hazardous. Though direct contact with ozone over a prolonged period can be harmful, as a purifier it acts rapidly to destroy germs and is then converted into other compounds, such as water, oxygen, and peroxides.

Ozone has been used extensively for water purification in France since the beginning of the century. It destroys bacteria and viruses much more quickly and effectively than chlorine. In potent concentrations chlorine may take about three hours to neutralize polio viruses, whereas ozone only takes about two minutes. Ozone is equally effective with other chlorine-resistant organisms such as *Giardia*, mycobacteria, Coxsackie virus, and encephalomyelitis virus, as well as with certain bacterial spores and amoebic cysts.

Ozonization is frequently used by commercial bottled water companies and is fairly effective as a means of obtaining purified water. However, it is most often used as a finishing process following either distillation or deionization. It is rather impractical for home use because of the high cost of buying and operating the equipment, but it is recommended where possible.

Ultraviolet Irradiation

Sunlight is our main source of ultraviolet irradiation. For centuries no one realized that the ultraviolet portion of the solar spectrum serves as a natural purifier of the environment, cleansing and disinfecting the surface waters of the world.

Today there are artificial sources of ultraviolet irradiation that not only mimic the sun but can also concentrate dosages that work much more quickly than natural sunlight. Ultimately ultraviolet irradiation could be the solution to purifying our public water supplies. For the present it can certainly be used for private supplies at home, in conjunction with distillation.

Because ultraviolet irradiation produces high-energy free radicals, direct exposure to it is dangerous. However, it has numerous effects that are helpful in water purification—disinfection, oxidative degradation, bleaching, clarification, deodorization, dehalogenation of carbon and nitrogen halogen compounds, detoxification of carcinogens and nerve poisons, formation of filterable precipitates, enhancement of biodegradability, decomposition of ozone-resistant heavy metal complexes, and elimination of bacteria, viruses, phages, protozoans, and algae.

Steam Distillation

When water is heated and allowed to boil, it changes to a gas vapor that rises, leaving behind the nongaseous impurities. When the gas cools, it changes back to liquid water that is free of all dissolved and undissolved nonvolatile substances. This is nature's principal method of recycling and purifying water. Ocean and surface water is vaporized by the heat of the sun; these vapors rise until they lose their heat energy, forming clouds that ultimately return the water to earth in the form of rain or snow.

Distillation removes 100% of all organisms, including salmonella and Legionnaire's bacteria, polio and hepatitis viruses, *Giardia*, amoebic protozoans, cysts, and pyrogenic substances. Water subjected to distillation does not need pretreatment except in very unusual circumstances, such as with raw sewage or water severely contaminated by petroleum distillates. Very hard water may require softening pretreatment. Distillation works well with seawater, brackish water, and highly mineralized water. It adequately eliminates chlorine, though the total elimination of chloramine requires additional ultraviolet irradiation.

Distillation, like other purification methods, can have drawbacks. If the feed water contains any volatile substances (such as chloroform) that have a lower boiling point than water, their gases can condense with the steam and end up in the final product. It is vital that the distiller have internal baffles (metal tubes within the boiler) to allow for fractional distillation, which permits the volatile gases to boil off first.

The beauty of distillation as opposed to any filtration technique, such as reverse osmosis or carbon filtration, is that purity does not vary with usage but is maintained throughout the life of the distiller. Filtration does provide purity to begin with, but the filtering material may lose its efficiency even after short-term use.

A great deal of energy is needed to convert water from a liquid into a gas. For this reason, the energy costs of distillation must be considered along with the cost of the equipment. Although some distillers are manufactured to work with gas stoves and camping stoves, those that use electricity are the most convenient for city dwellers.

There is some inconvenience associated with many home distillers. Most are simple pot boilers that must be filled manually with feed water, and the product has to be collected with tubing and stored in glass jugs. The residual minerals have to be emptied after each batch has been boiled. The usual maximum daily production is about 5 gallons, but this is adequate for most home needs. Many stills require considerable water for cooling, which is a waste of precious resources unless the water can be saved and reused; an air cooled distiller is preferable. Of the steam distillation systems presently available, the best are those that use internal baffles. Systems that also use ultraviolet irradiation or ozonization will give you added assurance that the water has the utmost purity.

In addition to the simple pot boilers, there are automatic distillers that don't need to be filled manually. The costs may range from $300 or $400 for the simple boilers to $2,000 for the fully automated ones. Distillers vary in size; some take little space on a shelf, some need as much room as a standard water cooler—about 1 square foot. The automatic distillers require electricity, a source of tap water, and a drain area for the concentrated wastes collected after distilling.

I have been working on a research project with some associates to develop an automatic distiller, but it will not be ready for manufacture for several years. In the meantime, there are a number of excellent distillers available from various manufacturers. Consult your Yellow Pages under the heading "Water Filtration and Purification Equipment" or contact the companies listed below.

Pure Water, Inc.   Home distillers
3725 Touzalin Ave.
Lincoln, Nebraska 68507

Sears   Portable distillers
Nationwide

Waterwise   Home distillers
Box 45994
Centerhill, FL 34254

Sprout House   Portable distillers
Box 700V
Sheffield, MA 01257

Using Your Distilled Water

Tap water should be consumed internally only for survival when good water is not available and you are in danger of dehydration. Otherwise, it is for external use only—for bathing, laundry, and household cleaning. Even then, there may be risks associated with tap water. Recent evidence indicates that chemicals in the water are absorbed through the skin. Toxic trihalomethanes derived from chlorine in tap water are gasified and may be inhaled when you bathe or shower. Taking shorter, colder baths or showers and ventilating the bathroom afterwards will reduce these risks. You should also avoid overexposure to chlorinated pools and spas. Laundry bleaching should be done in well-ventilated areas.

Certain medical studies have shown that hard (mineralized) water helps protect against heart attacks, whereas soft (lowmineral) water is associated with a greater number of heart attacks. However, these studies are flawed in that they do not take into consideration that chlorine and heavy metals (from water mains and from surrounding soil when the pipes leak) dissolve more easily in softened water and thereby raise the level of xenobiotics. Pure distilled water, though soft, does not travel through water mains and so is free of that problem.

Contrary to popular belief, distilled water *will not* leach minerals from your bones. Distilled water becomes thoroughly mixed with digestive juices as soon as you drink it. Once it is absorbed, it is no longer capable of acting differently from other body fluids.

Since distilled water does not contain any residual chlorine to destroy bacteria, it should be prepared on a daily basis or refrigerated until used. Aerating the water by shaking will improve its taste. Your bottled or home-distilled water should be utilized for all eating or drinking purposes.

1. Use distilled water for cooking and food preparation. From a practical standpoint, tap water may be used for washing fruits and vegetables, but if it is known or suspected to be high in xenobiotics, use distilled water instead. *Never* wash vegetables in Clorox or other chemical bleaches. Bleach has been recommended by some as a way to remove pesticides, but it should not be used for this purpose or any other dealing with food. Chlorine is extremely toxic; it reacts with living substances, as well as food, to form chloroform and other cancer-causing agents—and it does not remove pesticides.
2. Use distilled water in the preparation of all beverages.
3. Use distilled water for preparing infant formulas and drinks. Never use tap water.
4. Use distilled water for making ice cubes. It is inadvisable to buy commercially prepared ice since its purity is questionable. (Use your judgment. Ice cubes for that occasional party should not be a major concern.)

BEVERAGES

At the same time that you are altering your water-consumption habits, you should consider your overall fluid intake and the other liquids you consume. Many beverages contain harmful toxins that contribute to xenobiotic activity, and you must be aware of these so you can avoid them.

There is some disagreement about the amount of fluid needed each day for optimum health. The usual recommendation calls for 4 to 8 glasses of water per day. However, this rule does not always apply to everyone. For example, athletes and physically active people may need far more water, particularly on a hot, dry day. Use this standard amount as a guideline to be adjusted to your own lifestyle and needs. And remember that all fluids, not only water, contribute to the requirement, as do most foods.

The safest of all liquids is pure water. Other fluids, even if they are "natural," may have detrimental effects on health. Many of the beverages we consume are stimulants or depressants, and for some people it may be a considerable hardship to give them up totally. If this is the case, it is recommended that you cut down on them gradually. For the confirmed addict, one cup of coffee, tea, beer, or wine a day (but *not* one of each) can be considered permissible. For those who have signs of health problems, all caffeinated and alcoholic beverages should definitely be cut out entirely until the problems clear.

The Stimulants

Coffee, tea, cocoa, and kola nuts are natural pesticides, effective in repelling, paralyzing, or killing many insects. The same chemicals that make them toxic to insects serve as stimulants in humans: such chemicals as methylxanthines, theobromine, theophylline, and caffeine, which cause rapid heartbeat, insomnia, high blood pressure, and stomach distress, in addition to the desired stimulation and relief of fatigue.

The caffeine in coffee and other beverages has been linked to a number of health problems, such as cancer of the bladder and pancreas and fibrocystic disease of the breast. Recent studies have also found that coffee drinking is associated with elevated levels of triglycerides and cholesterol, with a resulting greater risk of heart attacks. Panic reactions and acute anxiety attacks can be triggered by the use of any of the caffeine-based stimulants.

In addition to the naturally occurring toxins in coffee, there are now man-made toxins to contend with. Coffee, along with other foods imported from Third World countries, contains significant amounts of pesticides that are now banned in the United States, including aldrin, BHC, chlordane, DDT, and lindane.

Many people, aware of the hazards of coffee drinking, have turned to decaffeinated coffee and herbal teas as a substitute. However, caution is also necessary here. Decaffeinated coffee often contains pesticides and residual solvents, though the water-processed varieties may be acceptable. According to my research, some herbal teas are safe (such as chamomile, anise, lemon grass, and aloe vera), but many present a significant hazard. Some of them contain pyrrolizidine alkaloids, which are carcinogenic, teratogenic (tending to produce deformed offspring), mutagenic (capable of causing mutations or changes in gene structure, which can lead to congenital birth defects), and abortifacient (inducing miscarriages). These herbal teas have caused liver and lung damage of epidemic proportions in Jamaica and South Africa, and laboratory testing has linked them to liver cancer in animals. The following are among the herbs that contain pyrrolizidine alkaloids.

borage	gravel plant	fireweed
comfrey	tansy ragwort	liferoot plant
coltsfoot	lungwort	bonesel
gordolobo yerba	Russian comfrey	senecio

Some herbal teas also contain substances that cause an anticholinergic reaction—dryness of mucous tissues, including the eyes, nose, and mouth, dilation of pupils, and a slowing of the heart rate. Herbs can cause allergies. Ginseng, a very popular herbal tea, sometimes causes hemolytic anemia, diarrhea, skin eruptions, insomnia, mental confusion, and swollen breasts. The following are among the herbs with anticholinergic properties.

burdock	juniper	jimson weed
hydrangea	catnip	

Soft drinks containing such subtances as caffeine (colas), bleached sugar, phosphoric acid, artificial sweeteners, sodium benzoate as a preservative, or artificial colors and flavors, as well as water of dubious quality, should be avoided. Various mineral waters containing natural flavors appear to be appropriate beverages. Hansen's and Napa Naturals are two companies that prepare their drinks with good water and acceptable ingredients.

Recommended Beverages

You may feel that you are being deprived of all your favorite beverages, but give yourself time. The drinks recommended below for good health and longevity can become enjoyable after you adjust to your change of habits, especially when you begin to notice how vigorous and alive you feel without all the toxins you have been consuming.

distilled water
verified safe artesian or spring water
fresh fruit and vegetable juices squeezed at home or recon-
    stituted from frozen concentrate with distilled water
commercially prepared juices, preferably fresh, unfiltered,
    and in glass containers rather than cans or plastic, with no
    added salt, preservatives, or other chemicals
soft drinks prepared with good water, fresh fruit flavors, and
    no preservatives or artificial flavors or sweeteners

If you can rid yourself of the xenobiotic toxins in drinking water and beverages this week, you will be achieving an important first step toward good health. Chlorine and the other chemicals that are added to most drinking water, along with chemicals from agriculture and industry that find their way into our water supplies, are among the most serious toxins ingested by people in the developed nations.

UPDATE YOUR SHOPPING LIST

Add
    distilled water or artesian water
    pure fruit and vegetable juices or concentrate
    natural soft drinks (if you must have soft drinks)

Remove
    beer, wine, and liquor
    coffee
    some herbal teas
    cocoa and sugared drink mixes
    colas and other soft drinks
    fruit and vegetable juices containing additives

WHAT YOU HAVE ACHIEVED THIS WEEK

If you have stopped drinking . . .	*you are not getting these toxins.*
tap water	chlorine, trihalomethanes, lead, cadmium, organic pollutants
beer, wine, liquor	solvents such as alcohol, acetone, aldehydes, fusel oils
coffee, tea	caffeine, pesticides
the harmful herbal teas	various toxins, including pyrrolizidine alkaloids
cocoa, colas	caffeine, pesticides
some commercially prepared vegetable and fruit juices	salt, preservatives

Chapter 5

Week Two: Vitamins and Minerals

Projects:
1. Consult your doctor about taking megadoses of vitamins and minerals.
2. Decide on an appropriate combination of supplements.
3. Begin taking your supplements three times a day.

This week's project is relatively easy but very important. We will concern ourselves with adding various vitamins and minerals to our daily regimen. If you are already taking supplements, you should know why they are necessary and how much you need. Many people are already taking multivitamins but are not really certain why. In this chapter we will attempt to point out a few of the reasons.

VITAMINS AND MINERALS AS ANTIOXIDANTS

Most people are aware that vitamins prevent such deficiency diseases as scurvy, beriberi, night blindness, rickets, and pellagra. There is mounting evidence that many vitamins and minerals are also antioxidants, or free-radical scavengers, and that they may be antidotes to various heavy metals, radiation, tobacco smoke, and other poisons. I consider vitamins and minerals necessary for maintaining daily good health and counteracting many xenobiotic-related disorders. Orthomolecular psychiatrists believe that megadoses of vitamins are effective in treating many psychiatric conditions.

Logically, it would seem that there is really no need for vitamin supplements. Certainly, before the discovery of vitamins early in this century, people survived well and lived healthy lives—if they did not succumb to cholera or tuberculosis. However, that was before the era of pervasive xenobiotics, before our food sources were exposed to a host of man-made toxins, and before we began to manipulate food in ways that destroyed its natural vitamins. If these were ordinary times and we were living in a toxin-free world, vitamin supplements would probably not be necessary. Food obtained directly from nature would supply all our vitamin needs.

But realistically, the toxins are here to stay. Fortunately, antioxidants can partially neutralize many xenobiotics, thereby protecting and preserving the normal functioning of the cells and chemical processes in our bodies.

In addition to mitigating the effects of ozone and singlet oxygen (highly reactive free-radical forms of oxygen), antioxidants offer some protection from the halogens—chlorine, fluorine, bromine, and iodine—and the halogenated hydrocarbons. Vitamins and certain minerals are necessary for body enzymes to function. They strengthen the immune system by boosting the production of interferon, T-cells, B-cells, and antibodies. Antioxidants also protect against viruses and bacteria.

Through evolution mammals have developed various enzymes with antioxidant capabilities that protect them from the toxic effects of oxygen metabolism. These enzymes include glutathione peroxidase, ceruloplasmin, glutathione, glutathione-s-transferase, catalase, and superoxide dismutase.

Not all antioxidants are equally necessary in combating chemical toxins in the human body, and some can actually be counterproductive, especially in large doses. For example, the synthetically produced food additives butylated hydroxytoluene (BHT), butylated hydroxyanisole (BHA), dodecyl gallate, and the sulfiting agents are very powerful antioxidants. Laboratory experiments have shown that BHT and BHA are capable of extending the life expectancy of rats, mice, and other lower animals. BHT and BHA also have cancer-fighting properties. But at the same time, they are synthetic phenolic compounds—xenobiotics—and they can act to magnify the carcinogenic action of such chemicals as 2-aminoacetylfluorene, urethan, 3-methylcholanthrene, and nitrosodimethylamine.

Epidemiologic surveys provide data suggesting that vitamin A, beta-carotene, vitamin C, vitamin E, and the minerals selenium and zinc offer considerable protection against cancer, especially lung cancer. The consumption of foods rich in vitamin A and carotenes has been shown to be protective generally, and the consumption of foods rich in ascorbic acid (vitamin C) appears to diminish stomach and esophageal cancers. Tests on animals exposed to cancer-causing agents almost uniformly show that retinoids and carotenes play a role in decreasing the number of developing cancers.

At this point, we do not have sufficient proof to state unequivocally that vitamins will prevent all man-made diseases, but we do know that they are at least partially effective in preventing some of them. Studies are needed to determine whether they are effective against the others. However, if we demand scientific proof before starting the antioxidant supplements, we will have a long wait. Remember, life is short and science is slow.

Most antioxidants are safe, but in large doses a few can cause side effects. For this reason, anyone who uses them should be aware of both the potential benefits and the risks. I have listed many of the risks and benefits below, but not all of them, nor have I listed all potentially useful supplements; that is beyond the scope of this book. There are, however, many excellent books on vitamins and minerals, including the Heinz Company's *Nutritional Data, Modern Nutrition in Health and Disease* by Drs. Robert S. Goodhart and Maurice E. Shils, and *The Merck Manual of Diagnosis and Therapy*.

A REVIEW OF THE VITAMINS AND MINERALS

Beta-carotene and Vitamin A (Retinol)

Carotenes are orange-yellow to red hydrocarbons that occur as pigments in many plants and exist in three isometric forms. Beta-carotene is sometimes called provitamin A because it can be converted in the body into two vitamin A molecules.

For some time it has been known that vitamin A, or retinol, is needed for proper color vision, for prevention of night blindness, and for proper growth and maintenance of teeth, bones, glands, nails, and hair. Vitamin A is usually derived from animal products such as milk, egg yolks, and liver. But beta-carotene is primarily found in yellow, orange, or dark-green vegetables, such as carrots, yams, squash, corn, pumpkins, and spinach.

Both beta-carotene and vitamin A are antioxidants, and in recent years it has been discovered that they are useful in protecting against cancer. Beyond its ability to convert to vitamin A, beta-carotene is effective as an anticancer antioxidant because it is a very powerful neutralizer of singlet oxygen. Because vitamin A is fat-soluble, it can be stored within the body for long periods, which may enable it to offset the effects of toxins stored in the liver and in fatty tissue.

A soon-to-be released study from Harvard University will show that beta-carotene supplements given to 20,000 physicians over several years have been instrumental in decreasing their cancer rate. Retinoic acid, a derivative of vitamin A, suppresses oncogenes, which are capable of converting normal cells into cancer cells. Another vitamin A derivative, beta-all-trans-retinoic acid, prevents or inhibits the development of precervical cancers.

Vitamin A is helpful in preventing lung cancers among both smokers and nonsmokers. This may be at least partially because lack of retinol can lead to increased carcinogen binding to lung DNA. Vitamin A also strengthens the immune system, whereas a lack of vitamin A and beta-carotene can suppress the production of the important T-cells and antibodies. However, vitamin A is no "miracle drug" for cancer prevention. It takes two years for supplemental use of beta-carotene or vitamin A to produce any benefit in cancer prevention, and four years before maximum results can be achieved.

Care must be taken with vitamin A; it is one of the few vitamins that may be toxic in large doses. Symptoms of excess vitamin A include headache, dry skin, vomiting, and drowsiness. Too much beta-carotene can cause a harmless yellowing of the skin, which may in fact help to protect against the harmful effects of ultraviolet irradiation and retard the development of sunburn and skin cancers. Thus, we are not including vitamin A on our recommended list (see Table 5.1) but will rely on beta-carotene instead.

People who take the drug Accutane should temporarily stop taking vitamin A. The two act synergistically and increase the risk of elevated blood levels of cholesterol and triglycerides. Accutane, which is a derivative of vitamin A and is prescribed for acne, is a teratogen and can cause birth defects and increase the potential for developing cancer.

Vitamin A toxicity should not be a reason to avoid taking supplements. Most healthy adults can safely take 30,000 IU a day without any side effects. In daily doses of over 100,000 IU, vitamin A can cause *pseudotumor cerebri*, a condition that mimics a brain tumor; in some cases, however, adults have taken the vitamin in excess of 100,000 IU a day over long periods without overdose symptoms. Children taking over 50,000 IU a day have shown evidence of toxicity.

If you are taking too much vitamin A, stopping should clear the symptoms; there is usually no permanent damage.

Vitamin D (Cholecalciferol)

The major source of vitamin D is exposure of the skin to the ultraviolet rays of the sun. This vitamin, along with calcium and phosphorus, is necessary for the development and maintenance of the bones and teeth. It is important for the immune system because it maintains proper blood levels of calcium, which is so vital to many cellular functions. A deficiency of vitamin D can result in rickets.

Vitamin D has no known antioxidant effects, but recent studies have shown that vitamin D and calcium are capable of decreasing the risk of colorectal cancer and that vitamin D can inhibit the growth of human melanoma cells in test tubes.

B-Complex Vitamins

The B-complex vitamins contain substances useful primarily as coenzymes for the respiratory enzyme system and the synthesis of tissue. Since this group has antioxidant capability, it appears to offer some protection against modern toxins. In some tests, certain B-complex deficiencies have been linked to an increase in liver tumors in rats treated with chemical carcinogens.

Vitamin B-1 (thiamine), the first of the B vitamins isolated, is a sulfur-containing compound. It is necessary for conversion of choline to acetylcholine (acetylcholine deficiency is associated with Alzheimer's and other diseases), for the functioning of many enzyme systems of the body, for daily metabolism, and for meeting the body's energy requirements

The antioxidant capability of thiamine protects against acetaldehyde, a chemical breakdown product formed in the liver from alcohol and also found in smog and tobacco smoke. Deficiencies of thiamine can occur during pregnancy or due to smoking and alcoholism. Of course, the best-known thiamine deficiency is beriberi, a disease that develops when people eat mostly grains from which the outer thiamine-rich covering has been removed. Beriberi was common in the past in areas where polished rice was the staple food.

Vitamin B-2 (riboflavin) is needed to maintain body tissues and is used by certain enzyme systems to convert fats, carbohydrates, and proteins into usable energy. It also helps keep the eyes from developing cataracts. Its importance in counteracting the man-made diseases stems from its function in the manufacture of one of the body's natural antioxidants, reduced glutathione. Riboflavin deficiency can result in a decrease of white blood T-cells, which are needed to ward off cancer and reduce susceptibility to certain infections.

Para-aminobenzoic acid (PABA) is a membrane stabilizer and sun-blocking agent.

Vitamin B-6 (pyridoxine) is needed for the formation of red blood cells and to maintain the health of gums and teeth. As an enzyme cofactor for the utilization of proteins and amino acids, it is also important for growth, for the conversion of glycogen to glucose, and for the conversion of amino acids to neurotransmitters. The vitamin B-6 requirement increases during pregnancy and with the use of sulfiting agents in foods, birth control pills, tobacco, and alcohol.

Vitamin B-12, choline, L-methionine, and folic acid are lipotropic substances involved in utilizing and mobilizing fats. Deficiencies of these substances have been associated with increased colon and liver cancers in laboratory rats. They help to prevent cancer in women taking birth control pills; taking folic acid reduces the incidence of precancers of the cervix.

Vitamin B-12 and folic acid are required for DNA synthesis. B-12 and folic acid deficiencies—generally caused by poor absorption rather than an actual lack—result in megaloblastic or pernicious anemia. Vitamin B-12 is found in meat products and is not manufactured by plants. Vegetarian diets have often been criticized for this reason, but evidence of B-12 deficiencies in vegetarians is scant at best. Indeed, vitamin B-12 can be obtained from microorganisms that live on certain plants and in plant foods such as soybeans, soy sauce, tempeh, or kelp. Even certain human intestinal bacteria can manufacture the vitamin. It can also be obtained from B-complex vitamin supplements.

Vitamin B-5 (pantothenic acid) is an antioxidant and anti-stress vitamin and is an important requirement for fat synthesis. It is helpful for people who suffer from Reynaud's disease, a condition in which poor circulation causes the tips of the fingers to turn white and numb during cold weather.

Vitamin B-3 (niacin or niacinamide) is essential for cell metabolism and absorption of carbohydrates in its niacinamide (nicotinic acid amide) form. Lack of this vitamin or its precursor, the amino acid tryptophan, can cause pellagra, a disease that affects the entire body but most particularly the central nervous system, gastrointestinal tract, and skin. Excessive amounts of the niacin (nicotinic acid) form of the vitamin, when taken as a supplement, can cause an uncomfortable itching and flushing of the skin, but this decreases in intensity after several weeks of continuous use.

Studies have shown that niacin, but not niacinamide, can lower plasma cholesterol and thus protect against coronary heart disease. However, extremely high doses may be necessary, with the attendant uncomfortable side effects. Although the patients in the cholesterol-lowering experiments used 2 to 9 grams of niacin daily, I believe that considerably less may be needed if the low-toxin diet and exercise recommendations are followed. One way of minimizing the flush and itch is to use timed-release niacin supplements in 125-milligram increments once or twice daily until the symptoms have disappeared. You may then substitute the less expensive regular tablets or take a combination of the two varieties. Since the amount required to lower cholesterol may vary from person to person, frequent follow-up blood tests for cholesterol, perhaps every three months, may be necessary until the cholesterol has normalized. Niacin in large amounts should be avoided by anyone who has gout (or an elevated blood uric acid level), ulcers, diabetes, or abnormal liver tests, or who is taking medication for high blood pressure.

Vitamin C (Ascorbic Acid)

Traditionally, vitamin C has been used to prevent the deficiency disease scurvy, which was a serious disease in earlier times. Scurvy caused Vasco da Gama to lose 100 men out of a crew of 160 during his voyage around the Cape of Good Hope in 1497. However, vitamin C has a great many other helpful properties that were not known in the past.

Vitamin C prevents the formation of nitrosamine compounds and protects against bladder, gastric, and esophageal cancers. It is possible that vitamin C may help to inhibit the spread of cancer by neutralizing hyaluronidase, an enzyme made by cancer cells. Vitamin C has antioxidant properties that help counteract the effects of halogenated hydrocarbons, mercury, cadmium, excess chromium, X-rays, and ultraviolet irradiation.

Vitamin C also has properties useful against human T-cell lymphoma viruses; its deficiency causes a decrease in the T-cells needed by the immune system. People with viral infections, such as common colds, flu, pneumonia, hepatitis, and mononucleosis, can recover faster and with fewer complications if they take adequate amounts of vitamin C. By fostering collagen formation, vitamin C also helps injured athletes and surgical patients to heal faster. It can prevent cataract formation in the eyes, and it helps counteract the toxic effects of cigarette smoke and polluted air.

Taking certain drugs, such as oral contraceptives, cortisone, and tetracycline, increases the need for vitamin C. Smoking, alcoholism, and excessive use of aspirin also decrease vitamin C storage in the body.

Since ascorbic acid, like insulin, can lower blood sugar, diabetics should be aware of this when using it as a dietary supplement and should be prepared to decrease insulin intake. Also, precautions should be taken for some forms of medical testing, because ascorbic acid will alter test results for hidden bleeding in the stool and possibly for blood sugar.

The major natural sources of ascorbic acid include parsley, broccoli, brussels sprouts, horseradish, citrus fruits, cabbage, cauliflower, and potatoes. The most potent antioxidant of the C group is ascorbyl palmitate, a synthetically produced fat-soluble form of ascorbic acid. It stays in the body for longer periods of time and helps detoxify the fat storage areas.

While not truly vitamins, bioflavonoids such as rutin and hesperidin, which are found in the rinds of citrus fruit, are antioxidants that supplement and enhance the antioxidant capability of ascorbic acid. Vitamin C's effectiveness is also increased when it is used in conjunction with vitamins A and E and selenium.

Vitamin E (Tocopherols)

Vitamin E is a fat-soluble vitamin found naturally in oleaginous vegetable and seed oils. It is important for healthy reproductive tissue, for embryonic growth, and for preventing certain nutritional-related muscular and neurologic disorders. It has been shown to be an extremely effective antioxidant and is the principal antioxidant found in nature. Tocopherols provide protection from high-pressure oxygen, ozone, nitrogen dioxide, and other oxidants. They have been shown to protect animals from the constituents of smog and may also be a defense against cancer in humans.

The tocopherols function with other antioxidants, selenium, and sulfur amino acids. Along with ascorbic acid, tocopherols aid in preventing the formation of nitrosamines. Vitamin E protects the vital areas within our cells against the poisoning attack of carbon tetrachloride and other chlorinated hydrocarbons. It is very important for those who continue to use polyunsaturated oils in their diets, as it protects polyunsaturated fatty acids from the oxidative effects of free radicals.

A study has shown that vitamin E may afford some protection against fibrocystic disease of the breast, a condition that occurs in approximately 50% of all women and increases their risk of developing breast cancer. In another study, vitamin E was shown to raise levels of high-density lipoproteins, but the test group was small and the results require verification, as does the possibility that total cholesterol levels may be reduced by large doses of vitamin E.

Large doses of tocopherols are relatively safe. In 1975 a study by the National Institutes of Health found no ill effects in subjects who had been taking large doses of vitamin E over a three-year period. However, Dr. Wilfrid E. Shute, a pioneer in tocopherol investigation, has warned that vitamin E in large amounts may cause problems in a few specific situations. For example, it causes a rise in blood pressure in 33% of hypertensive people, and it has some undesired effects on people with rheumatic heart disease. Therefore, you should reduce your blood pressure by dietary management before using megadoses of tocopherols. Vitamin E intake should be started at a low level and then increased gradually, with blood pressure being monitored frequently.

Vitamin K

Vitamin K is essential for proper clotting of the blood. It is a fat-soluble vitamin that is manufactured largely by bacteria in the human intestinal tract. Liver disease and poor intestinal absorption of vitamin K can cause problems with the manufacture of proteins needed for blood clotting. Vitamin K has no known antioxidant properties.

Calcium

Calcium is absolutely essential for the growth and development of bones in children and for the prevention of osteoporosis, particularly in postmenopausal women. However, it is a myth that milk is a perfect source of calcium.

Because of a lactase deficiency, most people of the nonindustrialized world are unable to drink milk or eat dairy products without developing diarrhea or severe abdominal pains. The people of most Third World nations obtain their calcium from vegetable sources, including leafy green vegetables, and they usually have no problem with osteoporosis. Since dairy products are also associated with other health hazards (see Chapter 7), they are an unacceptable source of calcium.

Under ordinary circumstances, vegetable sources of calcium are adequate. Almost all diets contain calcium in amounts above the threshold of human needs; a diet lacking in calcium is virtually impossible to find. It is the high phosphorus content of meats and other high-protein foods that causes the loss of calcium from teeth and bones and the development of osteoporosis (and sometimes cataracts, arthritis, and wrinkling of the skin as well).

A low-protein diet is the key to keeping bones strong. For the few people who have been eating a low-protein diet since early in their lifetimes, calcium supplementation is not necessary. The majority of people, however, do not fall into this category and should take a calcium supplement.

Selenium

Selenium is a mineral found in various foods, usually plants, which do not require it but absorb it from the soil. The amount of selenium in plants is dependent on selenium levels in the soil where they are grown.

In 1973 a major scientific breakthrough occurred with the discovery that selenium is part of glutathione peroxidase, an antioxidant enzyme that destroys lipid peroxides. These are intermediate compounds developing during free-radical activity within the body. Selenium, therefore, is vital in protecting against this dangerous activity. It was later found to be an antioxidant itself, in addition to its role as part of an enzyme. It is more effective when used with tocopherols.

In the United States, the western states have a high selenium content in rock formations and soil, while the eastern states, particularly those of the coastal plains and the northeast, have extremely low levels of selenium. It has been suggested that lack of selenium in these areas contributes to high death rates from cancer, coronary heart disease, and stroke. Cancer death rates for 1986, by state (see Table 9.1), correspond strikingly with the amount of selenium in the soil, and presumably in the food, of each state.

Selenium appears to confer some degree of resistance to the poisonous effects of methylmercury, cadmium, and arsenic by reacting with them to form harmless metal complexes. There is also a correlation between high selenium levels in the retina and the prevention of night blindness. Kashan's disease is a selenium deficiency of the heart muscle. Several reports indicate that heart attack rates are higher where selenium levels in the soil are lower, just as with cancer death rates.

However, there are risks associated with ingestion of excessive selenium. It can cause "blind staggers" in animals, and large amounts of the mineral in the agricultural runoff of the Kesterson Reservoir in California's San Joaquin Valley have been implicated, along with nitrates and pesticides, in deaths and deformities of waterfowl and other animals.

Zinc

Zinc is another important mineral. Trace amounts of zinc are necessary for protein synthesis and for general growth and development. Zinc is an essential component of approximately 80 enzyme systems.

But zinc's most important function is as part of the immune system. It helps to increase the number of T-cells and T-suppressor cells, which aid in the prevention of cancer. Zinc acts as a shield against cadmium toxicity; deficiencies of zinc and calcium amplify susceptibility to lead poisoning.

Other Minerals

There are many other minerals that are necessary for overall health and should be part of any good preventive program. These include magnesium, chromium, iron, and phosphorus. Copper (in small amounts) has important antioxidant properties. Iodides are needed in small amounts for proper thyroid functioning. Magnesium supplements (along with calcium) are helpful in preventing osteoporosis.

Antioxidant Sulfhydryl-Containing Amino Acids: L-Cysteine and L-Glutathione

These amino acids are necessary for the functioning of various enzyme systems. L-glutathione is an important part of glutathione-s-transferase, in turn a component of the glutathione redox cycle. This cycle defends lung cells against oxidative injuries resulting from hydrogen peroxide, smog, bleomycin, asbestos, and anthracyclic drugs.

L-cysteine is an important antioxidant in its own right; it is also a component of reduced glutathione, a body enzymatic antioxidant. When taken during meals, it causes a significant increase in the body's absorption of the iron present in vegetables. In conjunction with ascorbic acid and dapsone, cysteine offers protection to various vital inhibitors of proteolytic activity, such as alpha-1-antitrypsin. The protection is effective against such oxidizing agents as ozone, industrial gases, and cigarette smoke, and can help prevent the free-radical attack that may be responsible for chronic bronchitis, cancer, and emphysema.

L-cysteine is also capable of acting against cross-linking of molecules (one of the effects of free radicals). Along with ascorbic acid and thiamine, it can scavenge superoxide and oxidants generated by the white blood cell myeloperoxidase/hydrogen peroxide/halide system, thereby strengthening the body's immune system.

ANTIOXIDANT SUPPLEMENTS AS PART OF THE TEN-POINT PROGRAM

Antioxidants and other vitamins and minerals are a vital part of the low-toxin program. The dosages I suggest in Table 5.1 are considerably higher than the RDAs (Recommended Dietary Allowances) established by the National Research Council of the National Academy of Sciences. It is not possible to give a blanket recommendation that applies to everyone; we are not all alike, and our requirements vary depending upon age, weight, overall health, and other factors. Moreover, the doses I recommend are subject to change in light of further research. You must judge for yourself whether to take the high doses or not, weighing the risks against the expected rewards.

When you have your medical checkup before beginning the program, discuss the use of vitamin and mineral supplements with your doctor. Show him or her Table 5.1 so that the recommendations can be evaluated in relation to your physical condition. A doctor who is knowledgeable about nutrition and the use of antioxidants will be able to give you the best advice. Unfortunately, many doctors still consider vitamin supplements controversial and reserve them for treatment of deficiency diseases or for use as placebos.

While the rest of the ten-point program concentrates on the avoidance of toxins, the antioxidants serve as your main resource for fighting the toxins you have already consumed and stored, as well as those you will inevitably consume in the future. However, do not assume that they give you sufficient defenses to continue the high-risk accumulation of toxins from animal food sources, smoking, drugs, alcohol, or even tap water. Vitamins and minerals can do only so much.

The recommendations in Table 5.1 represent a standard for a person of average height and weight who is in good health and is relatively active; they must be adjusted for each individual. It is almost unheard-of for anyone, with or without health problems, to have other than minor difficulties with supplements in the dosages I recommend. However, you should not raise these dosages without good reason; they are already considerably higher than the RDAs. And remember that some vitamins and minerals can cause physical discomfort, such as intestinal gas, when you first start taking them, or health complications if you take them in excessive amounts.

TABLE 5.1
Recommended Antioxidant Dosages

VITAMINS	Total Daily[a]	% U.S. RDA[b]
Lipo-soluble		
Beta-carotene	25,000 IU	500
Vitamin D-3	200 IU	50
B-Complex Factors		
Thiamine-HC1 (B-1)	100 mg	6,666
Riboflavin (B-2)	100 mg	5,882
Niacinamide	250 mg	1,250
Pyridoxine-HC1 (B-6)	100 mg	5,000
Cyanocobalamin (B-12)	100 mcg	1,667
Folic acid	400 mcg	100
Pantothenic acid	2,000 mg	20,000
Biotin	300 mcg	100
PABA (para-aminobenzoic acid)	25 mg	—
Choline	250 mg	—
Inositol	250 mg	—

TABLE 5.1 (Continued)
Recommended Antioxidant Dosages

VITAMINS	Total Daily[a]	% U.S. RDA[b]
Vitamin C (ascorbic acid)	1,500 mg	2,500
Ascorbyl palmitate	1,500 mg	2,500
Bioflavonoids	500 mg	—
Vitamin E (d-alpha tocopherol)	1,000 IU	3,333
MINERALS (Chelated)		
Selenium (amino acid chelate)	200 mcg	—
Zinc (amino acid chelate)	50 mg	333
Calcium	1,500 mg	150
Magnesium	1,500 mg	375
Copper (amino acid chelate)	6 mg	300
Manganese (amino acid chelate)	7.5 mg	—
Potassium iodide	150 mcg	100
Molybdenum	250 mcg	—
Chromium GTF factors	50 mcg	—
Iron (fumarate)	18 mg	100
AMINO ACID ANTIOXIDANTS		
L-cysteine	200 mg	—
L-glutathione	15 mg	—
L-methionine	50 mg	—
MARINE LIPID CONCENTRATE		
OMEGA-3 FATTY ACIDS[c]		
Eicosapentaenoic acid (EPA)	1,620 mg	—
Docosahexaenoic acid (DHA)	1,080 mg	—

[a]My recommendations apply to adults and children age 14 and over.

[b]RDAs apply to adults and children age 4 and older. Where no figure appears in this column, there is no established RDA.

[c]Marine lipids need not be taken if your serum cholesterol falls below 180 mg/dl.

Guidelines for the Use of Antioxidant Supplements

- Do not start taking all the recommended supplements on the same day. Begin with one or two and keep adding new ones daily.
- Take your supplements three times a day with meals (food helps absorption), or at the very least, twice daily. Follow the recommended dosages in Table 5.1. A once-daily dose does not give you adequate antioxidant protection throughout the day.
- Do not expect megavitamins to give you a "high" of feeling good or full of energy (they might, but not necessarily).
- Do not use megavitamins as an excuse for not giving up tobacco and alcohol (though they do help the smoker and drinker).
- Do not use vitamins as a substitute for food. They do not have the calories needed to sustain life and will not prevent starvation.
- Vitamins cannot make up for a poor diet or for eating junk food or animal foods.
- Do not use vitamins as a substitute for vegetables.
- Vitamins cannot compensate for lack of adequate sleep.
- Take your vitamins no later than dinnertime if you are susceptible to insomnia (which has occasionally been a problem for those who take supplements late at night).
- Synthetic and natural vitamins should behave equally well, all other factors being equal (including lack of synthetic fillers). One exception is the vitamin E group, which is more potent in the natural form.
- Vitamins should not be considered a cure-all for emotional stress but only a part of an overall program.
- Do not expect vitamins to "cure" the killer or autoimmune diseases. Though they are highly recommended for people with these diseases, they are suggested here primarily as part of a preventive program.

UPDATE YOUR SHOPPING LIST

Most people who are unfamiliar with vitamin supplements are completely baffled the first time they survey the shelves at their local drugstore or health food store. There are numerous brands, with varying doses or combinations of vitamins and minerals, and it is easy to be put off by this admittedly confusing array. However, choosing your supplements is relatively easy as long as you are sure what you want. Proprietors of health food stores can be very helpful; show them Table 5.1 and ask their advice. It is even possible to purchase supplements by mail or via a computer modem.

To get the vitamins and minerals recommended here in the amounts suggested, you will have to take several preparations. Most commercially available preparations are synthetic and are just as effective as the natural ones. In fact, the small amount of "natural" vitamins added to some preparations has no particular effect other than to raise the price.

If you would like additional information concerning antioxidants, you may contact me at the Medical Center for Health and Longevity, at the address given in Appendix B. You will be sent a copy of my newsletter, which contains information on my ongoing research, the latest news concerning antioxidants, and other pertinent information.

WHAT YOU HAVE ACHIEVED THIS WEEK

By adding megadoses of antioxidants to your daily diet, you have taken a vital step toward protecting yourself against the hazardous substances in your food and in the environment.

Chapter 6

Week Three:
Exercise and Stress,
Vegetables and Fruit

Projects:
1. Develop a daily exercise program.
2. Determine your acceptable stress level.
3. Increase vegetable and fruit intake.
4. Begin using up all dairy products and eggs in stock.

During the third week of the program, you will again be concentrating on giving your body what it needs to defend itself against the toxicity of xenobiotics. Those who are already on a good program of exercise will have no difficulty with this week's projects and may wish to include some of the assignments from Week Four. Those who are overweight or lead sedentary lives will need to expend considerable effort in starting to exercise.

EXERCISE

Before beginning this or any exercise program, it is important to consult your doctor to determine the precise exercise level that is right for you, considering your age, weight, physical strength and stamina, and condition of health. A very high percentage of men and women in the industrialized nations are not in good physical condition and will have to start an exercise program gradually. This was not the case before the invention and widespread use of the automobile. In past centuries, people *had* to walk because they had no alternative; and so, for the most part, they were physically fit and did not need a calculated exercise program.

Fortunately, in recent years, a great many people have become aware of the importance of exercise and have begun to correct their old bad habits. However, they are still in the minority, and others need to join their ranks. Without exercise, muscles wither away and health suffers. Exercise contributes greatly to a sense of well-being, a positive mental attitude, and a capability of dealing with stress.

Exercise also lowers blood pressure and raises levels of high-density lipoproteins, which protect against coronary heart disease. Aerobic exercise helps the heart action become efficient, lowering the resting pulse rate. Physical conditioning helps prevent dangerous rapid heart rates during stress, so that shoveling snow or running to catch a plane or a burglar, say, is less likely to cause a heart attack. One theory holds that exercise increases the diameter of coronary vessels and improves the condition of the small capillaries, thereby improving circulation.

Exercise and Diet

It is important to realize that exercise alone will not resolve all health problems. Toxins in food are very closely linked to heart trouble, and exercise must be taken up in conjunction with dietary changes. The human body is not quite like a furnace in which all foods burn up completely like ordinary fuel. Toxic, fat-soluble chemicals become stored in the tissue no matter how active a person may be. Many athletes have developed coronary heart disease despite participating in active sports; running marathons does not automatically confer immunity from heart attacks or sudden death. It is important to combine your exercise program with the low-toxin diet in order to avoid heart difficulties.

A study performed by Swedish professor Per Olaf Astrand underlines the relationship between various foods and exercise. Nine men (three groups of three) were fed different diets for a period of three days, after which Astrand had them pedal a stationary bicycle until they were exhausted. Those who had been fed a meat diet that was high in fat and protein were able to cycle an average of 57 minutes. Those on a mixed diet of proteins, fat, and carbohydrates were exhausted after an average of 1 hour 54 minutes. The ones with the best endurance were those who had been fed a vegetarian, complex-carbohydrate diet; they were able to pedal for an average of 2 hours 47 minutes, almost three times as long as those on the high-fat, high-protein diet.

This experiment vividly illustrates how high levels of toxins in meat can diminish an athlete's performance on a short-term basis. Not only are these high-fat, toxin-laden foods inappropriate if you are trying to avoid the killer diseases, they also can worsen your day-to-day health and fitness. Long-distance runners take heed. Although you run, you may still have health problems if you don't restrict your intake of animal foods. The typical American diet is risky even if you run or exercise regularly.

Exercise and Weight

Exercise is of special concern to those who are overweight. Being overweight can decrease longevity. It can aggravate the onset and progression of many of the man-made diseases. And it can modify your quality of life by interfering with social and economic pursuits. For those who are overweight, moderately or severely, the low-toxin diet and exercise program can provide the necessary weight-loss (see Table 6.1) on a gradual but permanent basis.

To maintain proper weight, calories consumed must match the energy requirements of the body. Weight gain is caused by consumption of too many calories. And, of course, for weight loss to occur, you must consume fewer calories than the body requires. This is true for any weight-loss program, including the low-toxin program. In order to combine weight loss with reduced toxin intake and health maintenance, you should get your calories primarily from complex carbohydrates—fruits, vegetables, and whole grains. These complex carbohydrates will provide more than enough food to satisfy hunger while allowing a gradual, healthy loss of weight.

TABLE 6.1
Recommended Maximum Weights for Men and Women

Women			Men	
Height			*Height*	*Weight (Lbs.)*[a]
4'11"	91		5'2"	110
5'	94		5'3"	115
5'1"	97		5'4"	120
5'2"	100		5'5"	125
5'3"	104		5'6"	130
5'4"	108		5'7"	135
5'5"	112		5'8"	140
5'6"	117		5'9"	145
5'7"	122		5'10"	150
5'8"	127		5'11"	155
5'9"	132		6'	160
5'10"	137		6'1"	165
5'11"	142		6'2"	170
6'	147		6'3"	175
			6'4"	180
			6'5"	185

Note: Reprinted by permission of the Walter Kempner Foundation from *Bulletin of the Walter Kempner Foundation* (June 1972): 47.

[a]Figures are for weight fully dressed. Ideally, weight fully dressed should be below these maximums.

Exercise is also necessary as a means of burning off excess stored fat. Almost all weight-loss programs are more effective when combined with exercise. For every mile you walk or jog, you burn 100 calories. You must use 4,000 calories to shed 1 pound. A pound may not seem like much, but over time the weight loss from exercise and diet becomes meaningful. Simply walking 25 miles a week may result in a loss of half a pound, in addition to the weight loss from the low-toxin diet. Calorie counting is virtually unnecessary with this program.

First Step: Walking

Anyone who has been sedentary or whose cholesterol level is above 200 mg/dl should start exercise with a degree of caution. Narrowing of the coronary arteries is very common in a large segment of the population, and this is one reason it is important to have your heart status and cholesterol checked by your doctor before beginning the program.

It is not advisable for someone who has been sedentary to start right into a strenuous exercise program. It may have taken a long time to get your body into the shape it's in; don't expect to correct years of harm overnight. In the beginning, the goal is merely to reawaken stiff muscles and bring back a bit of the joy of living.

If you are a sedentary person, your first step toward exercise is to increase the amount of walking you do. Most inactive people walk 1 or 2 miles a day in the ordinary course of shopping or doing chores. Regardless of the extent of your present activity, make your eventual walking goal 5 miles a day, and build up to it gradually over a 6-month period. Begin this week, and stick with it consistently from this point on as a regular daily activity.

As you increase the distance you walk, you will need a good pair of running or jogging shoes. Don't delay beginning your walking program until you make this purchase, however; for a day or two, street shoes or tennis shoes will do. When you make your purchase, remember that your running shoe size will differ from your dress shoe size. Make sure you get shoes that fit and are comfortable. Nike, New Balance, Adidas, and Reebok are all very good.

If you can walk easily when you start, 15 minutes for the first day should be adequate. (If the weather is bad, you can always do your walking indoors.) If you have no major health impairments, increase your walking time by 5 minutes each subsequent day until you are walking for 1 hour a day. This should generally take no more than 2 weeks to achieve, but it may vary in individual cases. If it takes you longer, don't be concerned; set your own pace.

Once you reach 1 hour 15 minutes, you may stop adding time; maintain that level. If you wish, divide the time into two periods a day. At this point, assuming an average pace of 1 mile per 15 minutes, you should be walking 5 miles a day. And you should be feeling healthier and more vigorous than before.

Second Step: Aerobic Exercise

Although walking is ideal for reawakening muscles that have not been used in a long time, it is not enough to meet all your body's needs. Eventually you must work up to a program that will benefit your heart and lungs and improve your circulation. The heart has to be trained so that in times of stress, whether physical or emotional, it will not suddenly fail because of blocked circulation. The exercises that provide this training must make the heart and lungs work hard enough for you to perspire. However, I must again caution you that vigorous exercise may not be completely safe if your blood cholesterol level is above 200 mg/dl. If it is and you still wish to exercise, do so in a carefully monitored environment.

Aerobic exercise—exercise that accelerates the pulse rate— will benefit your cardiovascular system. It includes any form of exercise that uses large muscle groups (the arms, legs, and trunk). However, it is important that the exercise sustain rapid heartbeat. Exercise that stops and starts—tennis, handball, racquetball, baseball, basketball, etc.—is potentially hazardous to an untrained individual and should be avoided. These sports are for those who are healthy and have already developed good cardiovascular reserves. Also, such exercises as weightlifting, Nautilus training, or push-ups and pull-ups, which build muscle strength and bulk, are not designed to benefit the cardiovascular system, so they will not be considered for our purposes.

The exercises that best fit the needs of the ten-point program include running, brisk walking, bicycling, swimming, jumping rope, and aerobic dancing. The eventual goal is to maintain this type of exercise for an hour at a stretch.

Just as with walking, you should begin aerobic exercise gradually. Your choice of exercise should of course be made in consultation with your physician. Since individual tastes and needs vary widely, this volume cannot possibly cover all the exercise alternatives. You may want to consult one of the many books on aerobic exercise or enroll in a commercial program. Whatever you decide, make sure to follow a sensible program that is carefully planned to suit your needs.

The objective of aerobic exercise is to improve the circulation of blood through the heart and coronary arteries, and to make the heart muscle and lungs work more efficiently. The average normal resting pulse is about 72, but it may be higher for those who are in poor physical condition. With increased exercise, heart efficiency improves so that over time the resting pulse rate may decrease to 50 or 60.

While you are in the early stages of your aerobic exercise program, you should monitor your pulse. Use the formula of 220 minus your age to determine the maximum allowable pulse rate. In the beginning, limit yourself to 60% of the maximum, but go up to 80% as soon as you feel safe and comfortable with the strenuousness of the program (see Table 6.2).

If your choice of aerobic exercise is running or jogging, begin with a short run and increase the time by a few minutes a day, until you are running for a full hour; your ultimate goal is 5 miles a day 5 days a week. After each run or jog, check your pulse rate immediately to make sure it is within the 60% to 80% training range. If it goes over, you are exercising too hard; slow down a bit. Once you are conditioned, there is no need to keep checking your pulse.

TABLE 6.2
Pulse Rates During Exercise

Age	Maximum Heart Rate	Training Pulse		90% Rate
		60% Rate	80% Rate	
20	200	120	160	180
25	195	117	156	175
30	190	114	152	171
35	185	111	148	166
40	180	108	144	162
45	175	105	140	157
50	170	102	136	153
55	165	99	132	149
60	160	96	128	144
65	155	93	124	140

STRESS

Stress has been the subject of much study and discussion in recent years. Literally tens of thousands of articles have been written on the subject. Yet the concept of stress as a condition affecting human beings was not introduced until 1950, by Dr. Hans Selye, who borrowed the term from physics.

Most people today are familiar with the concept of stress and are able to recognize the effects of stress in their daily lives. We each experience different kinds of stress, and we react or respond differently. Even the definition may differ from person to person. For example, riding a rollercoaster may prompt exhilaration in one person while arousing mortal fear in another.

Some people perceive stress only as an emotional or psychological problem, closely linked with anxiety or the inability to cope with such matters as the death of a loved one, divorce, an IRS audit, severe illness, loss of job, a lawsuit, or other crises. However, stress goes beyond the emotional and includes any event that disrupts normal body functioning. It may involve anything from fever and infection to fear and anger, from the ingestion of low-grade poisons to severe injury or major surgery.

Dr. Selye defined stress as "the nonspecific response of the body to any demand." He made a distinction between pleasure and pain as different forms of stress, suggesting that the more we learn about the pain of living, the easier it becomes to enjoy the pleasures.

Dr. Selye's distinction may suggest a bit of good advice for us in attempting to deal with stress. Certainly we could never derive pleasure from some forms of stress, such as disease, grief, divorce, or injury. However, we can attempt to approach them with a positive mental attitude, and with other, more ordinary stresses, such as pressures at work, problems with children, or disagreements with family and friends, we can try to achieve some degree of satisfaction in being able to cope with them well. In many cases, the simple recognition that a situation is stressful can be of some benefit. Recognition may be the first step toward successful resolution.

Numerous methods have been devised to help people cope with stress. Among the techniques that have achieved popularity in the 1970s and 1980s are transcendental meditation, yoga, Zen, relaxation techniques, positive imagery, biofeedback, and behavioral modification programs such as Earth-Born, EST, the Esalen movement, and Arica, to mention only a few.

It is important to remember that different people respond to stress in different ways; some programs are beneficial to some people but not to others. Any program that helps you to approach stress in a positive way is right for you. In fact, this factor seems to be the universal one: the development of a positive mental attitude appears to be helpful for anyone.

Norman Cousins has suggested that a positive mental attitude is important in the treatment of cancer patients. The extent of possible benefits, however, remains a subject of controversy. A University of Pennsylvania Cancer Center study suggested that the cancer death rate was unaffected by emotions, lifestyle, or attitudes. It implied that patients were deceiving themselves if they thought that determination, confidence, hopefulness, and a strong will to live would play a role in their survival. Mr. Cousins concluded otherwise, arguing that all available means should be used to conquer cancer, including the emotional and psychological. He pointed out that the high cancer rate at that institution was not altered by the use of chemotherapy, radiation, and surgery, yet no one would suggest withholding those forms of treatment.

We are not attempting to cure cancer in this volume; we are merely offering counsel and trying to prevent cancer and other serious diseases. And stress can play a part in illness. Some have suggested that stress itself is not bad for us, that what is bad is our negative way of coping with it. There may be some truth in this. For this reason, part of your assignment this week is to begin to deal with stress in a more positive manner.

Recommendations for Dealing with Stress

- Make a list of the everyday stresses you experience. Analyze how you cope with them. Do you face them in a positive manner? Do you ever see humor in these situations? Are you able to laugh often throughout the day?
- Are you currently facing any serious stresses—divorce, illness, death in the family, fear of losing your job, and so on? Is there anything positive you can do to face these stresses? If you can do nothing, do you accept that fact or do you try to fight back anyway?
- Do you have ways of expressing your fears, anxieties, frustrations? Someone to talk to? Do you bottle everything up inside, or do you let it out? Are you willing to say to others, "I'm sorry, I can't deal with that right now"?
- If you realize that you deal with stressful situations in a negative manner, begin to think of ways to approach them more positively. When possible, try to see the humor in situations.
- Do you get enough sleep to be able to cope well during the course of an ordinary day? If not, attempt to get more sleep or better sleep.
- If you determine that you cannot cope with your stress problems on your own, consider seeking out a behavior modification program.
- Have you considered that an exercise program and a low-

toxin diet may help you to cope with emotional stress? There is reason to believe that inadequate exercise and synthetic chemicals in food can affect your attitudes and emotions.

VEGETABLES, FRUITS, AND GRAINS

This week you will begin to alter your food consumption. The change will take place gradually to allow your taste buds—and you—time to adjust. No one can be expected to give up lifelong eating habits overnight.

You will not be asked to give up eggs and milk products this week, but you will be asked to remove them from your shopping list in preparation for next week. Begin to use up all eggs and milk products you presently have in the house, and do not restock them. You should also begin to limit the amount of meat you eat while increasing the amount of fruits, grains, and vegetables.

Since you are beginning your exercise program this week, it will be especially helpful to add more carbohydrates to your diet. Carbohydrates, both simple and complex, should represent the largest percentage of your diet. They provide nutrients essential to most of your body's functions, from the thinking process to the workings of the muscular system.

Simple sugars, or simple carbohydrates, contain only one or a few molecules of such substances as sucrose, fructose, glucose, lactose, and maltose. They are found in foods like honey, granulated sugar, fruit juices, and syrups, which are often produced by means of processing techniques that separate the sugars from plant fiber. This fiber-free sweetener is absorbed quite rapidly into the body, causing an unnecessarily steep rise and then drop in blood sugar along with a rapid rise and fall of blood insulin levels. Fresh and dried fruits are the best sources of simple carbohydrates because they have sufficient fiber to help modulate these dangerous swings of blood sugar and insulin.

Complex carbohydrates are composed of strings of simple sugar molecules bound together to form long strands. During the process of digestion, enzymes such as amylase in the saliva and small intestine break the chains of sugar molecules into their simple sugar components. Vegetables and starch-rich foods—such as rice, wheat, rye, oats, potatoes, yams, sweet potatoes, beans, and peas—are excellent sources of complex carbohydrates.

Grains

Grains may come processed in the form of breads, pasta, noodles, and cereals. However, in supplying our complex carbohydrate needs, we want to avoid grains that have been bleached or more than lightly processed.

Grains should make up an integral part of your low-toxin diet, and your shopping list should include these items.

brown rice	wild rice	bulgur wheat
cracked wheat	barley	buckwheat (kasha)
millet	triticale	groats (whole oats)
rye	wheat berries	

In selecting rice, long-grain brown rice is recommended for easier cooking. Also, note that wild rice is actually a grass, not a rice, but its nutritional content is similar.

BREADS

It is important to obtain bread products made with unbleached flours, no sugar, and no hydrogenated oils. The words "chlorinated" or "brominated" in connection with flour should be a tip to stay away from the product. Unfortunately, salt seems to be a standard ingredient in breads, so it is virtually impossible to avoid. Recommended breads—but read the labels first—are as follows.

French	Italian	sourdough
pita	whole wheat	tortillas
matzo	bagels	whole grain crackers

One fairly reliable brand of sourdough bread is Pioneer. In selecting wheat bread, look for 100% stone-ground wheat bread or sprouted grain bread. If obtainable, Essene brand is good.

Tortillas should be made with corn or whole wheat flour, not bleached flour, without preservatives, and should not be fried. Matzo should be made with whole wheat flour or unbleached flour, preferably unsalted and without eggs. Whole grain crackers should not contain sugar, salt, or preservatives.

PASTA

Pasta is an important option in planning low-toxin meals, since 100% whole wheat pasta (as well as other whole wheat products) contains ample fiber. Of course, the pasta must be made with acceptable flour, not the bleached variety. Semolina flour or 100% unbleached durum wheat flour makes excellent pasta. Various vegetable pastas are also acceptable as long as they do not contain bleached flour, eggs, or artificial colors.

Many people still believe the myth that starches are fattening. They are not. (However, eaten in excess, any food is fattening.) Once you cut down on fats and proteins, weight loss is relatively easy on a predominantly carbohydrate diet, especially if exercise becomes a part of your lifestyle.

Vegetables

The dark-green and yellow vegetables—squash, pumpkin, carrots, yams, sweet potatoes, some melons, spinach, mustard greens, turnip greens, parsley, asparagus, okra, peppers, cucumbers, and others—are also a very important part of the low-toxin diet, not only for their nutritional value (large amounts of beta-carotene, ascorbic acid, and calcium, among other nutrients), but because plants are at the low end of the food chain and do not biomagnify the toxins present in the soil as do animal sources of food. In addition, carrots have been shown to lower blood cholesterol because they contain fiber and calcium pectinate, so if you have a cholesterol problem, eating two or three carrots a day seems advisable.

The cruciferous vegetables—cabbage, cauliflower, brussels sprouts, and broccoli—are particularly useful in avoiding cancer because they contain indoles (sulfuric antioxidant compounds). These should appear on the menu frequently.

Of course, the objective is to consume these vegetables in their fresh, natural state rather than canned, frozen, or processed. They all contain nutrients that are important in avoiding the major diseases. While fresh vegetables will inevitably contain some toxins from fertilizers and pesticides, we must attempt to keep the contaminants to a minimum by avoiding the additional chemicals used in processing.

Other vegetables and legumes on the unlimited usage list are the following.

potatoes	peas	beets
celeriac	celery	onions
chayote squash	chili peppers	garlic
collard greens	leeks	kohlrabi
daikon	artichokes	rutabaga
jicama	radishes	kale
Chinese cabbage	eggplant	parsnips
shiitake mushrooms	bok choy	snow peas
horseradish	Swiss chard	lettuce
tomatoes	cucumber	beans

Dr. James W. Anderson has demonstrated that soluble fiber like that found in legumes (peas and beans) and oat bran can lower cholesterol and raise HDL cholesterol. Accordingly, if you have a cholesterol problem, you should eat green and yellow split peas, lentils, black-eyed peas, and lima, navy, northern, fava, garbanzo, or pinto beans regularly (but don't overdo it if you want to lose weight, since they are high in calories).

Fruits

Fruits are high in water content, simple sugars, fiber, and many vitamins and minerals. They are a valuable, usually low-toxin food. They should be an integral part of your daily diet and should primarily be eaten fresh and raw.

You will want to include fruit in your meal plans not only for breakfast but for dessert at lunch or dinner. Fruit frequently contains one or more pesticides, but your body should be able to handle them as long as you are reducing your toxin intake from other sources. If your fruits are homegrown or come from reliable small farmers, the chances of their being relatively chemical-free are greater. Also, tropical fruit grown in Puerto Rico or Hawaii is free of many chlorocarbon substances found in fruit grown in Central or South America, though it may contain other substances. Unfortunately, even bananas, my favorite fruit, are shipped into the United States with an application of an antifungal and worm poison known as thiobendazole.

The following is a partial list of acceptable fruits.

apples	apricots	kiwis
pears	peaches	papayas
plums	persimmons	nectarines
pomegranates	oranges	pineapples
grapefruit	grapes	kumquats
mangos	tangerines	melons

Try to locate a supplier whose fruit is wax-free. The wax glistening on your apple may be the same type (carnauba wax) used to shine your shoes and your car, and it may also contain pesticides. Sometimes wax coatings are made from lard—certainly not acceptable on the low-toxin program.

"Vegetable Days"

One way of easing yourself into a change of dietary habit is to designate one, two, or three specific days this week as "vegetable days." On these days, plan to have a dinner meal consisting of steamed vegetables and a dry baked potato or brown rice, omitting beef, chicken, fish, cheese, and other animal products. You may add a tomato or lentil sauce to the potato in place of butter, cheese, or sour cream. For dessert, have fresh fruit.

If this is too much of a hardship for you this soon, allow yourself only a small amount of meat and decrease the portion sizes on subsequent vegetable days. For breakfast you may have several fruits or a bowl of steel-cut oatmeal. Lunch should be meatless; have either a salad, more fruit, or a bowl of minestrone soup.

Do not rely totally on your vegetable days to take care of your goal of increasing your intake of fruit, grains, and vegetables. You should concentrate on eating more of these every day this week.

UPDATE YOUR SHOPPING LIST

This week, in addition to removing eggs and dairy products from your shopping list, you should be adding a full range of grains, vegetables, and fruit. Use the recommendations in this chapter, along with your own preferences, as a guideline.

The fresher and more natural your fruits and vegetables, the greater their nutritional value. Of course, the only way to be sure that no pesticides or fertilizers were used in producing the fruits and vegetables you eat is to grow them at home. Even then, there is no way of knowing what contaminants are in the soil of your own yard or garden.

This is not to suggest that you should only shop in "natural food" or "health food" stores. If it were possible to be sure that the foods sold in such stores are indeed natural and healthy, that would be the ideal solution. Unfortunately, the data on this score are inconclusive. Some so-called natural or organic foods may be no less contaminated than those in your supermarket; others may be of significantly better quality. The best you can do is talk to your local store owners, find out how much they know about the sources of their produce, and use your judgment.

The federal government has failed miserably to provide us with adequate protection against pesticides and other xeno-biotics in our food. We as private citizens should support farmers who grow foods "naturally," and encourage other farmers to change. We should consider setting up private testing laboratories to check for toxins in our food, rather than relying on the government. The United Farm Workers of America, for example, has established a laboratory devoted to examining and reporting on California table grapes—a regular source of many deadly poisons. Entrepreneurs might consider that there is a viable business opportunity in the evaluation of foods for wholesale and retail establishments and the public; most people would probably be willing to pay a premium for fruits and vegetables known to be free of xenobiotics.

When possible, fruits and vegetables should come from growers who don't use pesticides. During the winter season in the United States, from November through May, this is usually not possible, because most produce comes from Mexico, Central America, and South America, where highly toxic pesticides are used more indiscriminately than in the U.S. A certification program for imported foods is especially important.

Farmers' markets, which exist in many rural areas and have begun to spring up in cities, give you a chance to speak directly to the growers and ask them if their produce is freshly picked and chemical-free. Will they be truthful? There is no guarantee, but the chances for pesticide-free fruits and vegetables are greater here than in a supermarket. If a farmers' market does not exist in your area, you may want to contact your local government to see if one can be organized. Many local governments encourage farmers to deal directly with the consumer.

Since toxins penetrate through the outer layer of vegetables and fruits, intense washing and peeling will remove only the superficial contaminants (and remember, *never* wash produce in laundry bleach). That's why it's important to buy your fruits and vegetables from as pure a source as possible.

WHAT YOU HAVE ACHIEVED THIS WEEK

If you have started . . .	*you are getting these benefits.*
walking	improved muscle tone (after some initial aches and pains)
doing aerobic exercises	decreased risk of heart attack, improved circulation, improved fitness, a feeling of well-being
working on stress	a positive mental attitude, a greater ability to handle emotional problems, more enjoyment from living
increasing complex carbohydrate, fruit, and vegetable intake	fiber and vital antioxidant vitamins, better bowel function, decreasing risk of serious illness, weight loss

Chapter 7

Week Four:
Breakfast and
Program Maintenance

Projects:
1. Stop using meat, eggs, and dairy products at breakfast.
2. Maintain your exercise program.
3. Continue increased grain, vegetable, and fruit consumption.

The major objective of the fourth week is to get accustomed to starting each day with a nourishing and healthy breakfast of fruit or whole grain cereals. This means phasing out the kinds of breakfasts a great many people today are used to having—if they have breakfast at all. No more eggs with bacon, ham, or sausage; no more donuts and coffee; no more heavily sugared rolls; and no crackles, puffs, pops, or crunches that masquerade as cereals but are in reality little more than processed sugar and salt. Most of these foods are poor in nutrition and high in additives. This week you will be replacing them with low-toxin foods that have the added advantage of supplying the nutrients needed to combat toxins.

The fourth week can be a crucial time in your low-toxin program. Only one major alteration has been scheduled for you this week in order to make sure you maintain the changes already made in your regimen. You should be firm in your resolve to consume only pure distilled water and beverages that are made with it. By now it should also be second nature to take your antioxidants regularly, three times a day.

It is equally important to maintain your exercise program without straying from your purpose. If you find yourself slipping or losing some of your resolve, go back and review Week Three. Remember to increase your exercise gradually to move forward toward your goal.

LOW-TOXIN BREAKFASTS

Whether your previous preference was for a "hearty" breakfast or a "sweet" breakfast, you will find that some adaptation is necessary to satisfy the requirements of the low-toxin diet. Your breakfasts should revolve around four basic foods: a variety of fresh fruits, oat bran cereal, steel-cut oatmeal, and cracked wheat cereal. Other whole grain cereals, whole grain pancakes, and breads made of unbleached or whole grain flours can be used for additional variety. Even such commercial dry cereals as Shredded Wheat, Grape-Nuts, or Nutrigrain wheat flakes are permissible. However, instead of putting milk on your cereal, substitute pure water, fresh or reconstituted fruit juice, or a mixture of the two. It may sound awful, but it's not. Because of our long-standing cultural bias in favor of milk, using water and fruit juice on dry cereal is not readily accepted, but it is actually quite enjoyable. Be patient while your tastes are changing. Remember, butter, margarine, salt, and milk are to be avoided.

In selecting breakfast cereals, the cooked whole grain cereals are preferable because they have undergone little processing.

Most of the following contain few or no preservatives or added chemicals.

cracked wheat	whole or cracked rye
wheat berries	cornmeal grits
Wheatena	triticale
groats (whole oats)	Zoom wheat cereal
steel-cut or Scotch-style oats	Roman Meal 4-grain cereal
rolled oats	oat bran

Some of these cereals—cracked wheat and wheat berries, steel-cut or Scotch-style oats—are available only at health food stores. For the most part, dry cold cereals are to be avoided because they contain excessive amounts of sugar, salt, partially hydrogenated oils, and other additives.

Bran, especially oat bran, can help lower cholesterol. If you have a cholesterol problem (180 mg/dl or over), steel-cut oatmeal or oat bran cereal should be your breakfast *daily*, with or without fresh fruit. You can also sprinkle about two tablespoons of oat bran into your soup or salad at lunch and dinner, or have oat bran muffins as your source of bran (see Chapter 15 for recipes).

The use of fresh fruit and fruit juice will enable you to vary your breakfasts and make them enjoyable. Fruit should, of course, be fresh rather than prepared or processed. Raisins or prunes that are sun-dried and do not contain sulfites or other preservatives may be used. If juices are made from concentrate, distilled water should be used to reconstitute them, as discussed in Chapter 4. Fruit juice, however, is high in concentrated calories. If you have a weight problem, avoid fruit juice until your weight becomes normal. Fresh fruit is acceptable if you are overweight, but eat only one or two fruits a day if you have elevated triglycerides, since fruit can raise triglyceride levels in sensitive persons.

Although fruit is essential in a low-toxin diet, many fruit products are too heavily adulterated to be of much value, and some have additives or preservatives that are actually harmful. Unfortunately, this is true of most jams, jellies, and preserves, especially those produced commercially. Because they are high in salt, processed sugar, and other chemical additives. you must remove most of them from your shopping list and your diet.

Homemade jams, jellies, and preserves can easily be prepared with proper sweetening agents and fresh lemon, without the use of preservatives, artificial colors, and meat-based gelatin thickeners.

You will be removing a number of items from your grocery list this week, among them the various breakfast meats you have been accustomed to: bacon, sausage, ham, lox, etc. Some of these are among the most toxic items on your shopping list, since they are heavily processed and loaded with fats, cholesterol, salt, nitrates, and nitrites. It is very important that you discontinue their use.

The Dangers of Dairy Products

Dairy products and eggs come from animals and thus stand very high on the food chain. Environmental contaminants concentrate in these foods, which makes them unacceptable for a low-toxin diet.

Dairy products have been associated with various allergies of the skin and respiratory system (ears, nose, throat, and lungs) and with behavioral problems. Infants may develop iron deficiency anemias when milk is the major part of their diet. Coronary artery disease, which is very common in industrialized countries, has been partially associated with high intake of milk starting at a young age. Recently, pediatric groups have recommended reversing this national habit. At various times such diseases as multiple sclerosis, ulcerative colitis, Hodgkin's disease, and Crohn's disease have also been linked to dairy products.

It is well known that antibiotic-resistant diseases can be spread through the meat or milk of cows given antibiotics either to treat infection or to stimulate growth. Antibiotic-tainted milk can also cause allergic reactions in sensitive individuals.

Milk and cheese products have been responsible for the transmission of listeriosis, an occasionally fatal disease caused by the bacterium *Listeria monocytogenes*. Adequate pasteurization of milk does not always destroy this bacterium. Other harmful organisms found in milk and dairy products include *E. coli*, staphylococcus, salmonella, and leukemia viruses.

Milk is often considered the perfect source of calcium for building strong teeth and bones or preventing osteoporosis. However, as explained in Chapter 5, most people outside the developed world have a lactase deficiency that does not permit them to tolerate lactose in milk, so they don't drink it. Thus, the majority of the world's population must obtain calcium from other sources, usually vegetables. They do not develop osteoporosis, which suggests that the calcium available from vegetable sources is adequate. Milk may actually be partially responsible for osteoporosis because of its high protein content.

If you removed milk, butter, cheese, and eggs from your shopping list last week and endeavored to use up what you had in stock, you should have very little of these products left. Use the remainder this week, but avoid them at breakfast. With time, you will see that they are not even essential for your lunch or dinner menus, and eventually you should no longer miss them.

UPDATE YOUR SHOPPING LIST

Add

steel-cut oats	whole grain flour
oat bran	whole wheat bread
cracked wheat	Kashi (a commercial mixture of 7 whole grains)
other cereals	

Remove

bacon	butter
Canadian bacon	cheeses
sausage	yogurt
breakfast ham	ice cream
breakfast steak	eggs
breakfast fish	processed jams, jellies, preserves
milk	
cream	

WHAT YOU HAVE ACHIEVED THIS WEEK

If you have stopped eating . . .	*you are not getting these toxins.*
bacon, sausage, breakfast ham, breakfast fish	salt, toxins in fat, cholesterol, nitrites, and nitrates
breakfast steak, eggs, milk, butter, cheese	toxins in fat, cholesterol
processed cereals	sugar, salt, additives
breads, pancakes, and waffles made with bleached flour	chlorocarbon residuals
jams, jellies, and preserves made with bleached sugar or preservatives	assorted xenobiotics

Chapter **8**

Week Five: Eliminating Processed Foods

Projects:
1. Begin to eliminate processed foods from your kitchen.
2. Remove processed foods from your shopping list.
3. Learn to read food labels.
4. Plan a low-toxin party.

This week you will be attempting to eliminate one of the major sources of toxins in your diet—processed foods. This will require considerable concentration and effort, for processing has insidiously affected almost all our food without our being fully aware of it. If you find that the suggested projects are too much to accomplish in a week, allow yourself two weeks. The extra time will pay off in the long run.

FOOD ADDITIVES

Food is processed commercially to achieve certain goals: to prevent bacterial or mold growth, to prolong shelf life, to add or remove color, to enhance flavor, or even to change one kind of food into something entirely different—to "create" a new food, primarily for economic reasons. During processing (milling, bleaching, etc.), foods sometimes lose some of their nutrients. We are interested in all forms of processing because, even though they may have a perceived benefit, they can affect us adversely.

Ever since humans developed agriculture and stable communities, there has been a need to store and preserve foods for barren periods. To survive, people had to develop methods of drying and smoking; later they learned to use oils, salts, and acids as preservatives. Much later, people began to alter foods for other reasons—taste, variety, and appearance.

Manipulating foods, for whatever reason, inevitably exacts a price. Although we do not know all the effects of food processing on our ancestors, we are aware of at least some damage. For example, removing the outer covering of grains and consuming preserved meats without fresh foods caused man-made deficiency diseases like scurvy, beriberi, and pellagra.

Some of the old techniques of preserving and altering foods have survived into the modern era, accepted without question because of their long history, despite the discovery of the dangers associated with them. Smoking and adding salt, nitrates, and nitrites are ancient methods still used today, though all have been linked to serious problems. Additives such as salt, nitrates, and nitrites contribute to high blood pressure, birth defects, cancer, and coronary heart disease. It has been discovered in recent years that the traditional Chinese practice of eating salted dried fish, which is high in nitrosamines, is responsible for a long history of cancer of the nose and throat. Smoking of food forms benzo[a]pyrenes, which are known carcinogens. Charcoal broiling of meat also adds benzo[a]pyrenes, as well as other polycyclic hydrocarbons.

With the coming of the Industrial Revolution, new food processing methods were devised, from heat sterilization and the evaporation of milk to the addition of new chemical preservatives (sodium benzoate, boric acid, sodium borate, salicylic acid, etc.) and synthetic colors. At the same time, food suppliers were becoming expert, hardened adulterers. Friedrich A. Accum, a German-born chemist, pointed out that the food supply early in the nineteenth century was being poisoned to an alarming degree. Alum was added to bread; red lead was used to color cheese; cream was thickened with arrowroot and flour; candy was brilliantly colored with copper and lead salts, which can cause chronic disease and death. The adulterers also cut the grade of tea by adding blackberry leaves. Capsicum was added to mustard. The poisons vitriol and *Cocculus indicus* were added to diluted beer and ale, giving them a nice bitter taste and a real wallop—and sometimes causing paralysis, convulsions, gastrointestinal problems, even death.

The discovery of aniline dyes from the distillation of coal tars added a new, very risky group of chemicals to our foods. Although the practice of coloring food to improve its attractiveness dates back to the ancient Egyptians and Greeks, their dyes were derived mostly from natural vegetable sources and were harmless. The new chemical dyes were foolishly thought to be harmless as well, so they were added to food supplies with little thought, and often with government cooperation; for example, in 1886 the U.S. Congress passed a law allowing the addition of yellow dyes to butter. By 1900, synthetic dyes were used on an ever-increasing number of foods, to the point where consumers would refuse to purchase many products, such as milk and butter, without the coloring agents. However, synthetic coal-tar dyes are xenobiotics and known carcinogens.

The twentieth century brought with it food processing on a vast industrialized scale, but with the emphasis shifted from preservation to cosmetic appearance, flavoring, and added shelf life. Though health is now a secondary concern at best, many manufacturers utilize advertising techniques to make the consumer believe their products are indeed healthy as well as attractive and tasty, the primary goal being increased sales.

With the development of food processing as a major industry, a whole new class of additives has become a part of our food chain. There are now over 5,000 chemical compounds available for uses as additives, with new ones being devised each year.

Some are harmless or relatively beneficial, such as vitamins, trace minerals, and natural mold retardants like calcium propionate. (However, vitamins and minerals are generally added merely to replace the natural nutrients removed by the food processor in the first place.) Other additives have no real nutritional value but serve as antioxidants, stabilizers, emulsifiers, thickening agents, preservatives, flavorings, colorings, or sweeteners. Most of these chemicals are xenobiotics.

The use of additives has received the blessing of governmental monitoring agencies in the United States. Frequently, because of intense lobbying by the food industry, even substances proved to be dangerous are not banned as food additives. And even when chemicals *are* proscribed by law, the food industry often manages to continue using them for years while lawsuits or appeals are pending in the courts.

But, one may ask, aren't tests done beforehand to determine the safety of chemicals in foods? Yes, but the test results are not always accurate, for a variety of reasons. Some high-risk substances pass because the tested doses have no effects on the limited number of animals in the study. Others have no harmful effects on animals but prove harmful to humans because of the difference between animal and human enzymes.

One very serious flaw in testing techniques is that additives are tested one at a time and are not required to be tested in combination with other chemicals. Thus, additives may be acting chemically or synergistically with other additives in the same product or other products, but the tests will not disclose this danger.

Another problem is that of "dosage." The acceptable daily intake, or tolerance level, of an additive is the theoretical amount that can be taken for a prolonged period, presumably a lifetime, without damage. Tolerance levels are based on animal experiments, with a "safety factor" calculated so that the data can be applied to humans. The problem is that very broad assumptions are made in determining safety and that no allowance is made for the interaction of additives or for the fact that people may consume more than one product with the same additive. And once a chemical is approved, the maximum dose is frequently exceeded by the food processor, either accidentally or intentionally.

However, even low doses of some chemicals have been known to cause a great deal of difficulty for certain individuals. For example, the milk of dairy cows treated with antibiotics such as tetracycline and penicillin can trigger an adverse reaction in a sensitive person. Tartrazine Yellow #5 and sodium benzoate can cause hyperactivity and allergic reactions in sensitive children and adults.

A major feature of the government's regulation of food additives is the GRAS list, an extensive list of additives that are "Generally Recognized As Safe" by the U.S. Food and Drug Administration. Additives in use prior to the 1958 passage of the Food Additives Amendment to the Food, Drug, and Cosmetic Act of 1938 were included on the list without testing if scientific consultants considered them safe. The GRAS list is periodically updated as new additives are tested and old ones now determined to be unsafe are dropped. However, it still contains numerous chemicals that have never been tested, many of which are clearly not safe, especially when they interact with other "safe" or unsafe chemicals. Even those that have passed the GRAS tests may pose hazards.

Table 8.1 lists the major food additives (not all on the GRAS list), their primary uses, and their potential problems. Other food additives on the GRAS list include the preservatives dilauryl thiodipropionate and ascorbic acid; the sequestrants citric acid, sodium hexametaphosphate, and calcium hexametaphosphate; the anticaking agents calcium silicate, sodium calcium aluminosilicate, and aluminum calcium silicate; the synthetic flavors acetaldehyde, acetoin, benzaldehyde, n-butyric acid, cinnamaldehyde, strawberry aldehyde, vanillin, ethylaldehyde, methyl anthranilate; vitamin supplements; spices, sugar, salt, pepper, monosodium glutamate, and vinegar; and the multipurpose items hydrochloric acid, phosphoric acid, caffeine, and magnesium hydroxide.

In addition, the FDA allows pesticides in food and sets tolerance levels for them. The poisons permitted include aldicarb,

TABLE 8.1
The Major Food Additives

Chemical	Use	Nutritional Value	Potential Problems
Nitrates and nitrites	Meat preservatives that prevent botulism	None	May be carcinogenic, may cause birth defects and behavior disorders
Salt, sugar, vinegar, pickling brine	Preservatives in meat, fish, vegetables	Little	May cause strokes, high blood pressure, stomach cancer
Benzoic acid, sodium benzoate	Preservatives in soft drinks, other foods	None	May be carcinogenic, may cause headaches
Butylated hydroxytoluene (BHT), butylated hydroxyanisole (BHA)	Preservatives and antioxidants in breakfast cereals, baked goods, oils, etc.	None	May cause behavior disorders, rashes, allergies
Sulfiting agents (sulfites, metabisulfites, bisulfites, sulfur dioxide, etc.)	Preservatives in salads, wine, medication, dry fruit	None	May cause asthma, rashes, intestinal pains, shock, death
FD&C reds, yellows (Tartrazine #5, Azo Red, etc.)	Synthetic colorings in cosmetics, candies, cakes, desserts, oranges, etc.	None	May be carcinogenic, may cause behavior disorders, asthma, rashes

(*continued*)

TABLE 8.1 (Continued)
The Major Food Additives

Chemical	Use	Nutritional Value	Potential Problems
Chlorine, iodine, and bromine compounds[a]	Bleaching agents in white flour, sugar, cornstarch, bleu cheese, cottage cheese, tripe	None	Chlorinated hydrocarbon residues that may cause cancer and behavior disorders
Hydrogenated and partially hydrogenated oils	Prolongs shelf life of fats in margarine, cereals, baked goods, bread	Limited	Linked to coronary heart disease, Crohn's disease

[a]Strictly speaking, these are not additives, but they are used to treat foods and do leave deposits.

benomyl, captan, 2,4-D, DDT and metabolites, diquat, malathion, paraquat, simazine, toxaphene, and oryzalin, among many others.

With the use of food additives so prevalent, and with government agencies supposedly monitoring our food for "acceptable" levels of chemicals, we tend to lose sight of the most important and obvious question: Is there a truly safe level for xenobiotics, especially those that are cancer-causing? Any sensible person will answer no.

Preservatives

Of all the additives, the preservatives have historically seemed the most beneficial. They have been responsible for the availability of food during droughts and crop failures and in the dormant season. Preservatives prevent the growth of bacteria and mold, thus protecting food from spoilage for long periods. Yet even the most ancient of food preservation techniques pose risks to health. Today refrigeration, freezing, and the ready availability of fresh foods through rapid transportation make chemical preservation less necessary and easier to avoid for those who wish to follow a low-toxin plan.

NITRATES AND NITRITES

Sodium and potassium nitrates and nitrites are used as preservatives, most commonly in the preparation of delicatessen meats, to protect against the growth of *Clostridium botulinum*, a bacterium that produces the toxin responsible for botulism food poisoning. Nitrates and nitrites also give an attractive bright red color to these meats, along with a distinctive flavor. Small amounts of nitrates are found naturally in carrots, radishes, and lettuce. Nitrates are also a contaminant in the drinking water of many areas of the country.

As beneficial as nitrates are, there are considerable risks associated with them. Bacteria cause nitrates to undergo a chemical conversion into nitrites, which in turn can be converted into nitrosamines in the gastrointestinal tract. Nitrosamines can cause birth defects and are responsible for the formation of stomach and nasopharyngeal cancers.

Heating foods by almost any means—barbecue, oven range, frying, toasting, roasting, or dehydrating—causes small amounts of nitrosamines to form whether nitrates are present or not. This includes such foods as bacon, luncheon meat, frankfurters, beer, toasted or malted grain products, broiled meats, dried powdered foods (such as nonfat milk, dehydrated vegetables, cheeses, or instant coffee).

Nitrosamines may appear in rubber products, including baby bottle nipples, especially if the nipples are heated along with the bottle contents. Migration of nitrosamines from paperboard packaging into food has been demonstrated under laboratory conditions.

SULFITING AGENTS

While the FDA dropped sulfite preservatives from the GRAS list in July 1986 and banned their use on fresh fruit and most vegetables in restaurant salad bars, sulfites are still found in a variety of other restaurant foods, processed foods, beverages, and drugs.

Sulfiting agents (sulfur dioxide, sodium sulfite, sodium and potassium metabisulfite, sodium and potassium bisulfite) are antioxidants and are used to preserve wine, beer, cider, fruit juice, lemon juice, some carbonated and other soft drinks, some packaged fresh mushrooms, syrup, maraschino cherries, mustard, jam, salad dressing, some fresh and frozen fish and seafood, processed meat and vegetable products, frozen potatoes, frozen dough products (including pizza), dried fruit, and canned soups, as well as many other canned, frozen, or dried food products. Not all foods containing sulfites have them listed among the ingredients.

Sulfiting agents can affect humans in a variety of ways. They are known to decrease the body's ability to absorb vitamin B-1, and they can cause asthma, severe itching, hives, swelling, major abdominal pain, headaches, and even shock and death. Smog contains sulfur dioxide, which poses a hazard to those who are sensitive. Some medications contain sulfites—even medications for treating asthma. Local anaesthetics may also contain sulfites; many people who at one time thought they were allergic to local anaesthetics used in dentistry have discovered that they were actually responding to the sulfites. Fortunately, as more drug companies have become aware of the sulfite issue, they have either reduced or eliminated the sulfites in many drug formulas.

Although avoidance is the best way to handle sulfites, taking large amounts of vitamin B-12 (1,000–5,000 mcg) by mouth partially blocks the asthma caused by sulfites. Sulfite-sensitive people, particularly those who frequently eat in restaurants, may be exposed without their knowledge; they may wish to consider purchasing dipstick test strips (available from Center Laboratories, 35 Channel Drive, Port Washington, NY 11050) to warn them of the presence of sulfites in food and beverages.

Bleaching Agents

Bleaching agents are used on foods primarily for cosmetic purposes. They also shorten the time needed for aging and "conditioning" flour. Some of the chemicals used for bleaching and conditioning are from the toxic chlorocarbon group and may include compounds containing the halogens (chlorine, bromine, iodine, fluorine), such as monochloramine, bichloramine, sulfochloramine, chlorine dioxide, potassium bromate, and potassium iodate.

Since bleaching may add serious toxins to foods, bleached foods should be avoided. Unfortunately, the fact that certain foods may be bleached is not always mentioned on their labels; such foods include bleu cheese, cottage cheese, tripe, flour, sugar, cornstarch, and vegetable oils. Assume that *all* flour, including "enriched" flour, is bleached unless it is whole grain or the label specifically states otherwise.

Cheeses such as bleu and Gorgonzola have traditionally been made from goat's or sheep's milk and are relatively white in appearance. However, when they are made from cow's milk, a yellow color develops. This presents a problem in selling them to the consumer, who expects these kinds of cheese to be white. The manufacturer's solution is to use a bleaching agent so that the cheese looks white the way it is "supposed" to. However, along with the white color come chlorinated hydrocarbon residues.

In its natural freshly milled state, wheat flour has a yellow tint because of the presence of carotenoids and other natural pigments. During storage, however, wheat flour slowly and naturally becomes whiter as it undergoes an aging process from oxidation. This same oxidative process also makes the flour more elastic, producing a stable dough of good bread-making quality.

Early in this century, it was discovered that flour could be aged and matured instantly by bleaching rather than waiting for nature to take its course. A very common bleaching agent used in the early years of the industry was agene, or nitrogen trichloride. However, in 1946 it was discovered that white bread made with agene-bleached flour caused dogs to have agitated fits and was probably unsafe, so other bleaching agents (also potentially unsafe) were substituted.

The benefits of bleaching agents are relatively small compared to the dangers. Bleaches not only remove valuable vitamins and amino acids from foods, but they also deposit small amounts of xenobiotics. In laboratory experiments, 66% of rats fed a diet of bleached "enriched" white bread died within three months, while the growth of the remaining rats was stunted. It is not unusual for insects to bypass bleached white flour and wisely head for unbleached whole wheat flour.

Flavor Enhancers and Dyes

The additives that make foods appear more attractive or flavorful without contributing nutritional value have been a subject of controversy for years. They have been considered safe in the past and were even listed on the government's GRAS list. Years later their true toxic nature became known. A few were subsequently removed from the GRAS list, but many continue to be used because of their commercial value.

SYNTHETIC FLAVORINGS

There are over 2,000 agents used to enhance the flavor of foods, and they are the most common food additives. While some of the flavorings are natural and relatively harmless, the majority are synthetic. Some synthetic flavors have chemical structures similar to the natural ones, but many are purely man-made. The advantage of the synthetics for food processors is that they are less expensive than natural flavorings and that they often add to the shelf life of the product.

For example, real chocolate has about 1,200 chemical components, 6 times as many as natural strawberry or lemon. In order to keep expenses down, many cocoa manufacturers have resorted to using synthetic chocolate flavorings instead of real chocolate. These concoctions have included such chemicals as pyrazines, phenylethylamine, trimethylamine, and others.

Whether the resulting flavors are realistic is debatable, and their safety has also been called into question.

MONOSODIUM GLUTAMATE

Monosodium glutamate (MSG) has little taste of its own but tends to intensify the flavor of other foods when it is used as a seasoning. MSG is the salt of a naturally occurring amino acid isolated from seaweed in 1912. It is also found in fermented soy products. Although MSG is supposed to be a flavor enhancer, people who do not use it for a while find that foods taste much better without it.

Monosodium glutamate is responsible for allergic symptoms, including asthma, headaches, facial flushing, and hives. It is a neuroexcitatory chemical and can cause hyperactivity in children. When researchers discovered that MSG caused chromosome, brain, and neurologic disorders in infant mice, it was removed from baby foods.

SYNTHETIC DYES

The artificial food colors have one of the worst safety records of all the additives. Of the 24 or more food dyes used since the beginning of the century, 17 have been banned from the market.

There are serious concerns about the use of dyes that convert foods into artificially bright garnishes. FD&C Red #3, used in the production of maraschino cherries, is known to cause thyroid cancers in laboratory animals. FD&C Yellow #5 and FD&C Yellow #6 are still being used in foods although both have caused various types of allergic reactions and are suspected carcinogens. All three dyes have caused chromosome damage in animals. Several other undesirable dyes are used primarily in cosmetics.

For whatever reason, the Food and Drug Administration has not seen fit to remove these dyes from the market even though there is good cause to do so. However, you can easily remove them from your diet. Do not buy any food that has artificial flavorings or colorings added.

Sweetening Agents

NATURAL SUGARS

Natural sweetening agents—glucose, fructose, honey, and sorbitol—are carbohydrates. All sugars are converted to glucose during the digestive process, and in normal amounts the body can make use of them without harm. But when natural sugars are bleached, serious risks can occur, so processed sugar should be avoided. As explained in Week Three, one nutritional advantage of natural sugars from fresh fruits and vegetables is that they come amply supplied with fiber, unlike honey and the processed sweeteners.

Even for people who are excessively overweight or whose blood cholesterol and triglyceride levels are high, natural sugars are preferable to synthetic sweeteners. These natural sugars do *not* include bleached white sugar, brown sugar, or colored white sugar. The recommended natural sugars, in moderate amounts, are raw unbleached beet or cane sugar, unsulphured blackstrap molasses, concentrated fruit juices, honey, and unprocessed syrups.

The synthetic sweeteners have no place on a low-toxin diet. The cyclamates, saccharine and the newer aspartame (Nutrasweet), are potentially mutagenic or carcinogenic agents.

SACCHARIN

Since its discovery in the late ninteenth century, saccharin has been used as a replacement for sugar. In the late 1960s, it was suspected of being a carcinogen, but its use was not limited in the United States until 1978, when tests showed it caused cancer in laboratory animals. There are still those who fight for the right to use it, even though the National Cancer Institute has cautioned that it should not be used by nondiabetic children and pregnant women, and has discouraged its use by anyone else.

Saccharin should be avoided by anyone following a low-toxin diet.

ASPARTAME

Since aspartame is so new, the controversies surrounding its use have not been settled. Discovered in 1965, it was not ready for the market until 1973. Even then, there were doubts about its safety, for it was found to break down into its constituent amino acids (phenylalanine and aspartic acid) when heated. If it is stored in warm areas or kept on store shelves for a prolonged period, aspartame will change to methanol, an alcohol that ultimately converts to formaldehyde and formic acid, known carcinogens.

Soon after aspartame went on the market, sensitive people who used it began to complain of dizziness, headaches, hives, and menstrual irregularities. Some researchers believe that aspartame can also cause behavioral changes, particularly in children, because of its high concentration of phenylalanine (harmful for those children with a condition known as phenylketonuria). The aspartic acid component of aspartame, with or without glutamate, is a neuroexcitatory agent capable of causing brain or neurologic damage. It also greatly enhances the neuroexcitatory capability of monosodium glutamate.

Aspartame is not recommended in a low-toxin diet.

Nonfoods

Some food products are so completely synthetic that they are in a class by themselves. Virtually all the processing additives are used in their manufacture—preservatives, colorings, flavor enhancers, and bleaching agents. We may consider these products "nonfoods," though some call them "junk foods" or "snack foods." All of them should be avoided, as they are almost completely made up of potential toxins.

The nonfoods include packaged cakes, cookies, candies, cheese snacks, party snacks, party dips, cheese spreads, soft drinks, powdered drink mixes, canned puddings, low-cholesterol egg substitutes, nondairy creamers, and such cereals as crackles, pops, puffs, etc.

FOOD LABELING AND PACKAGING

In the United States, food companies are required to list the contents of most packaged or processed foods on the labels. In some cases, these listings are reliable. FDA regulations require that the contents of processed foods be listed according to percentage, so that the ingredients first on the list constitute the highest percentages of the product, and so on in descending order (the actual percentages do not have to be given). This procedure was instituted (over the objections of the food companies) to inform consumers of the nutritional benefits and chemical risks of products. To some extent, that goal has been realized.

However, most people do not read the fine print on food labels but trust the merchandising claims printed much larger on the front of the package. Since consumers have become concerned about natural foods, the food manufacturers have begun to advertise "natural ingredients," "organically grown," "no preservatives," and "natural flavors," and most people believe these claims. Often there is little truth to them. To be able to say that a produce has "natural ingredients," all manufacturers have to do is add one or two, usually salt or sugar. There is no law to prevent them from making such absurd claims.

Unfortunately the terms "health food," "natural," and "organic" have no legal meaning. The food companies themselves can decide on their own definitions, which may or may not coincide with *Webster's*. Although Wisconsin cheddar cheese sold in health food stores can legitimately claim to be natural, Kraft uses the same term to promote Cheez Whiz and Velveeta, two foods so highly processed they do not even meet the federal government's definition of cheese.

The government has "standards of identity" that define what can and cannot be in a product labeled as a specific food. The objective is to prevent manufacturers from adulterating products (e.g., by grinding up old moldy cheese and repackaging it as "processed cheese," or adding cultures to spoiled milk to produce sour cream). However, these standards only pertain to a selected small group of processed foods, and the remainder go unregulated. Because of this loophole, unscrupulous manufacturers can use deception in labeling with little risk of interference.

The food laws have many other loopholes that benefit food processors. One of these permits some foods to be sold without ingredient labels. Staple foods made from a standard recipe, including ice cream, mayonnaise, ketchup, canned vegetables, milk, margarine, and some breads, do not have to list their ingredients. Whole wheat bread may be sold with 14 or more optional additives that do not have to be disclosed. In these cases, consumers have no choice but to avoid the products if they wish to minimize their consumption of toxins.

For our own protection, it is imperative that we begin to read the fine print on all food labels before purchasing the products. Ideally we should learn what each complicated chemical is, but that is impractical in a world where new ones are devised almost daily. A good rule of thumb is that if an ingredient is totally unrecognizable, it is almost certainly a man-made chemical and probably harmful or of doubtful safety.

Below are two examples of popular products found in your supermarket. A quick glance will reveal that there are few "natural" ingredients, and virtually none of value to anyone interested in a low-toxin diet.

Instant Vanilla Pudding
  white sugar
  dextrose (corn sugar)
  modified tapioca starch
  sodium phosphates (for proper set)
  salt
  hydrogenated soybean oil
  BHA (preservative)
  Di- and monoglycerides (to prevent foaming)
  artificial colorings (including Tartrazine Yellow #5)
  nonfat milk
  artificial flavoring
  natural flavoring

Deluxe Yellow Cake Mix
    white sugar
    enriched bleached flour (with iron and vitamins)
    vegetable shortening (partially hydrogenated soybean oil)
    leavening (baking soda, monocalcium phosphate)
    propylene glycol monoesters (for smooth texture)
    dextrose (corn sugar)
    salt
    polyglycerol esters (for smooth texture)
    cellulose gum (for smooth batter)
    xanthum gum (for smooth batter)
    ascorbic acid (to protect freshness)
    modified food starch
    soy protein isolate
    Synthetic coloring (FD&C Yellow #5)
    artificial coloring
    artificial flavoring

These are not isolated examples. A trip down the aisles of your supermarket will reveal countless products equally as synthetic. They should all be avoided.

Reading Pasta Labels

Since noodles and pasta are a vital part of a low-toxin diet, it is important to recognize which products should be avoided and which are acceptable. Many of the pasta products in the supermarket are unacceptable because they are made with bleached white flour. It is necessary to read the labels to see if the contents have been processed in this manner.

Refined products containing 100% unbleached durum wheat or semolina flour, with no eggs, hydrogenated oils, colors, or preservatives, are acceptable, and their taste is superior to that of traditional pasta. However, do not buy pasta simply because the package says "Made with Durum Wheat," for it may contain as little as 2% durum wheat flour. Spaghetti products made with 100% whole wheat flour are also acceptable.

Reading Bread Labels

Bread is another staple of the low-toxin diet, and it must be selected with care. Do not trust the word "pure" or "natural" on the front of a bread package without investigating the list of ingredients. Baking companies are no different from other food manufacturers in their tendency to exploit the public's desire for wholesome food by deceptively labeling their products "natural."

In selecting bread, you should never buy products made with bleached flour, even if the word "enriched" appears on the label. White breads are always made with bleached flour and therefore should always be avoided. However, it is important to realize that some supposedly whole grain breads also contain bleached flour and should also be avoided.

Look for breads whose labels indicate they are made with unbleached flour or 100% whole grain stone-ground flour, yeast, water, and little or no salt. Such sweeteners as fruit juice, raw sugar, and honey are acceptable, but not white or colored white sugars, excessive salt, or partially hydrogenated vegetable oils. Rye, whole wheat, sourdough, pita, French, and Italian breads with acceptable ingredients are commercially available.

If bread labels list unintelligible chemical names, you should know to avoid the products.

Reading Labels on Canned and Frozen Foods

Ideally you should avoid all preserved foods on a low-toxin diet, eating only fresh foods. However, the ideal is not always possible. When fresh fruits and vegetables are available, certainly they should be chosen over frozen or canned goods, but there are times when compromise is necessary.

The canning process gives foods a long shelf life but is not without its faults. Processors often add salt, sugar, partially hydrogenated oils, and various other surfactants, acidulants, preservatives, and coloring agents during canning. You must read a can's label to make sure a product does not contain any of these substances. Another problem with canned goods is that when they are stored for long periods of time, traces of lead, cadmium, and arsenic can leach from the seams of the cans into the food. The canning process also slightly increases the amount of mutagens in the food.

Frozen "gourmet" dinners and low-calorie "slimming" dinners are veritable chemical storehouses and should usually be avoided. As a rule frozen vegetables are acceptable, although some manufacturers use chelating agents to remove iron and zinc so the green color doesn't turn gray with time.

Although it is nearly impossible to avoid processed food completely, you should try to keep your intake of it to a minimum. The rewards to overall health are well worth the sacrifice in convenience.

Food Containers and Cookware

Very little can be done about the packages and containers foods are sold in. Nevertheless, you should be aware that even these can add small amounts of toxins to the foods you consume. Of all the possible food containers, the ones least likely to add contaminants are glass and ceramics, though some ceramics manufactured in Third World nations do contain lead in the glazes. Lead-free glass containers are the safest means of storing leftovers.

As mentioned before, the good old-fashioned tin can may sometimes leach arsenic or lead into the food. There is some linkage between aluminum cookware and Alzheimer's disease. In fact, in the case of aluminum, the evidence of toxicity is suggestive enough that it may be advisable to use other types of pots, pans, and wrapping material. Plastic, however, may be one of the worst offenders. Although "food grade" plastic is highly regarded by the food and bottled water industries, water and acidic foods can dissolve gases and chemicals from plastic, including methylene chloride (the same solvent used to remove caffeine from coffee). Some processors add BHT to cardboard boxes, which have been known to contribute small amounts of BHT to the food products they contain.

GIVING A LOW-TOXIN PARTY

Removing processed foods from *your* diet should not cause you concern about giving a party for your friends. It is important for you to realize that the low-toxin program does not mean you have to stop entertaining, and that avoiding processed foods does not mean the end of your social life. That's why one of your projects this week is planning a low-toxin party.

Remember that a party is *not* the place to convert your friends to a low-toxin lifestyle. Do not force your guests to eat exactly as you do. Of course, most of the food should be acceptable to you, but plan to provide a few of the snacks and beverages your guests will expect—nuts, chips, soft drinks, wine, beer, or liquor. In the case of soft drinks, you may choose to serve only those made with pure water, natural flavorings, no preservatives, and no artificial sweeteners. Prepare and serve some new alternative foods as snacks for yourself and your friends who already understand the value of low-toxin living (Chapter 15 contains recipes for salads, gazpacho, and other dishes you may want to adapt for party purposes); your other guests may try them and find them as enjoyable as the party food they are accustomed to.

For a post-jogging party brunch, concentrate on fresh fruits, juices, steel-cut oatmeal, and vegetables—tomatoes, sliced onions, carrots, celery, cauliflower, cucumbers—with avocado or fresh salsa dips.

Including "natural" foods at your party table should not make you feel insecure. Your guests and friends will respect your decision to remain healthy. In time, as they see the good results you have achieved, they may decide to avoid the processed food snacks at your next party.

UPDATE YOUR SHOPPING LIST

You will probably be using a lot of ink on your shopping list this week, crossing out processed foods. Most of the items on your original list should no longer be purchased or stocked. Examine the foods you are accustomed to buying and those you currently have on hand, as well as those on your shopping list, to determine if they contain or have been subjected to any of the following.

wood smoke
drying and/or salting
  (sodium chloride)
bleaching
salicylic acid or sodium
  salicylate
boric acid or sodium
  borate
calcium or sodium sorbate
nitrates and nitrites
sugar
vinegar, acetic acid,
  acetates, diacetates, or
  pickling brine
benzoic acid or sodium
  benzoate

butylated hydroxytoluene
  (BHT)
butylated hydroxyanisole
  (BHA)
sulfiting agents (sulfites,
  bisulfites, metabisulfites,
  sulfur dioxide)
ethyl and propyl parabens
  (p-hydroxybenzoate
  esters)
epoxides (ethylene and
  propylene oxides)
diethyl pyrocarbamate
antibiotics
hydrogenated or partially
  hydrogenated oils

This week you should be endeavoring to get rid of foods containing the above substances. These include delicatessen meats, hotdogs, bacon, sausage, prepared ham, most commercial salad dressings, canned meats, chips, "junk food," most commercially prepared cookies and cakes, some breads, cream- or cheese-based party dips, mayonnaise, TV dinners, and candies. Some condiments may be used in small quantities. Such products as pickles, olives, soy sauce, ketchup, mustard, and relishes are available without sodium benzoate or other preservatives, added sugar, or excess salt.

Also examine the food on your shelves and your shopping list for any of the following ingredients.

salt	N,N'-di-o-tolyethylene
white or granulated sugar	diamine
(bleached sugar)	FD&C Yellow #5
saccharin	FD&C Yellow #6
aspartame or Nutrasweet	FD&C Red #3
monosodium glutamate	FD&C Red #4
dioctyl sodium sulfosuccinate	FD&C Red #40
cyclamic acid	FD&C Blue #1
	FD&C Green #3

The number of foods containing these potentially harmful additives is so long we cannot list them all here. A partial list includes confections, food coatings, dessert powders (chocolate, vanilla, and butterscotch puddings), gelatin desserts, cake and cookie mixes, bakery goods, cake decorations, sugar wafers, chocolate cookies and cakes, very dark chocolate, ice cream, ice cream cones, sherbets, ice cream bar coatings, spray-dried cheese, American cheese, butter and margarine, some spices, candied dried fruits, maraschino cherries, powdered drinks, carbonated beverages, casings for sausages and hot dogs, cheese-flavored snacks, all candies (including hard and striped candies), chewing gum, medication (tablets, liquids, and capsules), pet foods, and frozen dinners. Remember that restaurant foods may also contain these additives. Sometimes synthetic dyes are used even on fresh fruits, vegetables, and meats.

You should remove all items containing any of the above toxins from your shopping list and finish using what you have in stock. In some cases, it may be worthwhile to toss them into the garbage, especially if any of the additives in the second group above are high on the list of ingredients.

WHAT YOU HAVE ACHIEVED THIS WEEK

If you have stopped using . . .	*you are no longer getting these toxins.*
luncheon meats, bacon	nitrates, nitrites, fats
frankfurters	animal protein, additives
frozen fish	animal protein, sulfiting agents
canned meat, canned fish	animal protein, additives
prepared cakes, cookies, and mixes	chemical colors and additives
white breads and other bleached flour products	chlorocarbon residuals
hard candy, chewing gum	bleached sugar, artificial colorings, flavorings, sweeteners
cheese- or cream-based party dips	animal fats, artificial colorings, preservatives, etc.
mayonnaise	vegetable oils, eggs, preservatives
maraschino cherries	artificial red dye
ice cream	animal fat, artificial colorings, artificial flavorings

Chapter 9

Week Six: Minimizing Environmental Hazards

Projects
1. Inspect your household for toxic chemicals.
2. Determine whether your home has proper ventilation.
3. Inspect your workplace for toxic chemicals.
4. Learn about your neighborhood.

There is unfortunately not a great deal we as individuals can do about toxic chemicals in the environment; this is a broad social concern and must be dealt with on a worldwide basis. However, there are certain steps you can take for your own personal protection. Week Six will be largely a process of becoming aware of and guarding against the poisons that are closest to home. Some of these, such as household and garden products, are within your control, and you can find alternatives to them. Toxins in the workplace may be impossible to avoid, but you can find ways to protect yourself from them. Be aware that certain chemicals in the workplace may cause problems

134   *Choose to Live*

even though their levels do not exceed the standards of the Occupational Safety and Health Administration (OSHA). If you live in an area where there are toxic waste sites, educate yourself and become more involved in the political process to try and eliminate them.[1]

HOUSEHOLD TOXINS

Whether you live in an apartment or a house, once you begin to inspect your home for potential toxins, you will be surprised at the number of dangerous products in your possession, many of them used on a daily basis. The modern "miracles" of chemistry appear in products for personal hygiene and grooming, cooking utensils, kitchen and bathroom cleaners, laundry products, fungicides and insecticides, paint and insulation, and gardening fertilizers. All are potentially harmful, and their contents should be investigated.

Personal-Hygiene and Grooming Products

Since the recent discovery that the skin is not a chemical-proof wrapping, it has become clear that many of the substances we apply externally are absorbed by our bodies. There are many recorded instances of soaps, toothpastes, deodorants, hair dyes, sanitary tampons, and cosmetics causing serious problems, often labeled "allergic reactions" but sometimes acknowledged as toxic reactions.

Virtually any preparation devised to alter or improve upon the body's natural chemical functions is suspect. Whenever a new soap, detergent, antiperspirant, or perfume hits the market, the business of the local allergist or skin specialist seems to thrive.

One of the earliest of the "allergic reactions" was associated with the antiseptic hexachlorophene, which had been used as

[1]The Center for Science in the Public Interest has compiled a good book on the environment and the toxins in it: A. J. Fritsch, ed., *The Household Pollutants Guide* (Garden City, N.Y.: Anchor Press/Doubleday, 1978).

a powerful germ killer and surgical scrub for many years. Some surgeons discovered very early that they could not use it because of the severe rash they developed. Yet it remained on the market and was used in the late 1950s and early 1960s in a variety of commercial products, most notably Phisohex and Dial soap.

As a result of exposure to hexachlorophene, many people developed long-term skin rashes as well as phototoxicity (which means that the rashes became worse when exposed to sunlight). It was discovered that hexachlorophene was absorbed through the skin, remained in the body for a long time, and could cause toxic neurologic symptoms and even death, particularly among premature infants washed in hexachlorophene products. With this discovery, hexachlorophene was removed from the market; Dial soap and other products no longer contain it.

Some of the synthetic coloring agents used in hair dyes and makeup are known to be carcinogenic in animals, yet they have not been banned from use in products for humans. Presently there is a lawsuit pending to force the FDA to outlaw various synthetic colors used in foods and cosmetics, but it has met with little success so far.

We do not have sufficient knowledge to be absolutely certain which of the personal hygiene and grooming products on the market today are entirely safe. However, we do know the dangers of some chemicals. Certainly any product whose label recommends taking an allergy test before using it, as some hair dyes do, should be avoided. In examining product labels, be particularly wary of chemicals whose names are unfamiliar. However, even familiar names are suspect if they are not chemicals found in nature but instead come from a scientist's research laboratory.

Read the labels of the products you use or are considering using, and avoid those that contain the following chemicals: halogenated hydrocarbon propellant or fluorocarbon aerosol; solvents such as alcohols, ethers, acetone, and terpenes; artificial colors such as Acid Black 52, FD&C Red #3, FD&C Red #4, and FD&C Yellow #5 (tartrazine); perfumes and fragrances; preservatives such as propyl or methyl parabens and sodium bisulfite; and additives such as ammonium hydroxide, ammonium acetate, hydroquinone, resorcinol, p-aminophenol, lead acetate, aluminum chlorhydrate, 5-bromo 5-nitro-1, 3 dioxane, and methylchloroisothiazolinone. Aluminum may be linked to Alzheimer's disease, and antiperspirants containing aluminum should be avoided; be sure to read the label.

Some of the above chemicals are known to cause cancer in laboratory animals; some do not. Allergic or toxic skin rashes, however, can be caused by these as well as other chemicals. As with food additives, if you do not recognize the name of a chemical in a grooming product, consider it potentially unsafe until proven otherwise. We simply do not know enough about the long-term effects of synthetic chemicals.

In recent years, a number of manufacturers have been marketing hygiene and grooming products made with "natural" or "hypoallergenic" ingredients. These do appear to be considerably safer than products heavily laden with chemicals, but be certain that the "natural" substances are the only ingredients and not just a tiny fraction of the contents. One product that appears to meet this requirement is Kiss My Face, an unscented pure olive oil soap available in health food stores.

Household Cleaning Products

In the twentieth century, we have become dependent upon chemical products that keep our homes clean and shining brightly without a great deal of effort. However, some of these miraculous aids that free us from scrubbing, scouring, and polishing contain toxic ingredients that can be absorbed through the skin or by breathing the fumes.

Floor cleaners and polishes, sink and tile cleaners, ammoniated products, air fresheners, furniture polishes, oven cleaners, drain solvents, window cleaners, laundry detergents, and fabric softeners should all be investigated for suspected toxins. You cannot be expected to give up all of these products, but you can take every possible precaution when using them; do not allow them to come into contact with your skin, and avoid inhaling their fumes. When in doubt, of course, use good old-fashioned soap and water.

Among the chemicals to be handled with extreme caution are household bleaches containing chlorine or hypochlorite; ammonia (*never* mix ammonia with household bleach, as deadly chlorine gas can be generated); sodium hydroxide (lye), used as drain or oven cleaner; and sulfuric, hydrochloric, and phosphoric acid, used as drain, toilet, or denture cleaner, and very dangerous.

Other Household Chemicals

Not all of the potentially hazardous chemicals and chemical products around your house are used for cleaning. Some products seem harmless because they are not in liquid form, but remember that supposedly inert plastics and polymers in *any* form, including clothing fabric, rugs, and upholstery, can easily generate toxic gases if exposed to flames or intense heat. That's why it's unwise to attempt to iron some of the modern no-iron fabrics at a high temperature. You should also be extremely cautious in using plastics or polymers in the kitchen, especially around the stove or oven. Saran Wrap and similar plastic wraps will form deadly dioxins if allowed to burn.

Teflon coating for pans is a great boon for those who wish to cook without fats, but it is important to know that when Teflon is accidentally heated above 280° it produces toxic fumes in the form of polytetrafluoroethylene. These fumes are potentially harmful to humans and in numerous instances have killed household birds and cats; signs of toxicity in cats have appeared within 30 seconds of exposure, and death can occur within 3 minutes. Something a simple as leaving a cigarette burning in a Teflon pan or allowing the pan to boil dry over a flame can cause the release of toxic fumes. If this happens, remain in the room only long enough to prevent a fire, then vacate the area until it has been thoroughly ventilated. Humans can suffer acute respiratory and flulike symptoms from exposure to these fumes.

The recommendation to avoid water stored in plastic holds

equally for food. Scientific reports indicate that water stored in plastic containers leaches methylene chloride from the plastic. Water and acidic foods are solvents, and it is likely that they can leach other chemicals from plastic containers. Your body does not need these toxic molecules. Use glass containers wherever possible.

Another important concern is proper ventilation or air circulation in homes. In recent years, because of efforts to cut down on energy consumption in houses and apartment buildings, there has been an emphasis on insulation of walls and ceilings. Two types of foam insulation widely used in home and mobile home construction—urea formaldehyde and polyurethane—do have toxic risks associated with them. The most serious risk is to workers during the manufacturing and curing process, for that is when the fumes are most acute. However, urea formaldehyde can release toxic fumes over a long period of time, so it can affect residents too. Both kinds of insulation will release toxic fumes if they are burned.

Fiberglass and asbestos have also been used as home insulation and may contribute to health problems. The risk of cancer from exposure to asbestos has been highly publicized, but it is not well known that fiberglass, which is used in drapery, sporting goods, and various other products in addition to insulation, also poses a risk. Animal experiments have suggested the possibility that fiberglass causes cancer but have not as yet provided conclusive evidence. However, whether it is carcinogenic or not, fiberglass remains a serious problem; it can cause skin rashes, asthma, bronchitis, sore throats, nasal allergies, sinusitis, laryngitis, and nosebleeds.

Some fresh air is important in homes and offices, both in winter and in summer. However, because of the need to conserve energy, many older buildings have been tightly insulated and provided with double windows. Newer buildings have been constructed as hermetically sealed airtight shells. "Fresh" air is provided through interconnected heating and air-conditioning systems. The tendency of modern architecture to seal buildings off, allowing only for recirculated heated or cooled air, presents a hazard. People who live or work in such buildings may experience a variety of symptoms, including fatigue, headaches, muscle aches, dizziness, sore throats, and asthma—a phenomenon known as the "tight building syndrome," or the "sick house syndrome," which is becoming more common. Buildings made of concrete blocks or poured concrete tend to emit radioactive radon and other toxic fumes. The fumes contain dust, hydrocarbons, carbon 7–11 alkanes, carbon dioxide, carbon monoxide, ethylbenzene, formaldehyde, hexane, nitrous oxides, ozone, perchloroethylene, toluene, trichloroethane, and xylene. Of course, smoking in sealed houses exacerbates this problem.

One chemical that tends to build up in enclosed homes and offices is formaldehyde, which is found in insulation, cosmetics, deodorants, toothpaste, medications, hair conditioners, paper products, no-carbon copying paper, permanent press clothing, book bindings, adhesives, and resins. It also gets into the air by way of cigarette smoke and incomplete combustion from gasoline and diesel engines. Even combustion from gas home heating and cooking produces small amounts of formaldehyde. From all these sources, fumes can build up inside tightly sealed structures.

Lawn and Garden Products

Among the most dangerous of the chemical products around the home are those devised for use in lawns and gardens. As mentioned earlier, fertilizers, insecticides, and fungicides are among the most serious toxic pollutants we have. These are used not only by large farmers and growers but by ordinary citizens who spend their weekends grooming lawns and gardens.

While many chemicals have been banned in the United States, there are still numerous potentially unsafe products on the market. It is recommended that you avoid using pesticides and fungicides on your lawns and gardens, especially if you raise fruits and vegetables, and that you limit or avoid the use of chemical fertilizers.

In addition, the empty containers should not be discarded

indiscriminately or via the usual trash collection method. Most municipalities do have special collection days for toxic wastes such as used pesticide or solvent containers and half-empty paint cans. I am not convinced, however, that such toxic wastes are properly disposed of by municipalities, or that there is indeed such a thing as "proper disposal."

As a concerned citizen, you should also question communitywide spraying programs to eliminate mosquitoes or fruit flies.

TOXINS IN THE WORKPLACE

It may not be quite as easy to identify toxins in your place of work as it is in your home. Certainly all potential hazards cannot be covered here, because each occupation involves different chemically produced products. However, some guidelines can be given to help you develop your own line of investigation. It is certainly not suggested that you change your occupation, for every line of work today may have its risks, even if you are a secretary or a clerk in a store.

The concern for construction, construction materials, and ventilation, mentioned in reference to your home or apartment, is even greater in the workplace. The accumulation of fumes from various departments of a company, or from various companies within a single building, may result from the use or presence of a wide variety of chemical products singly or in combination, especially if air is recirculated. It may be impossible for you to determine what is in the air you are breathing, and you cannot protect yourself against the unknown, except to get as much fresh air as possible.

What you can do something about are the chemical products you work with directly. In some cases you may be able to replace them with alternative products; in others you may keep their use to a minimum; and in very serious cases, you may choose to wear gloves or a protective mask to keep from touching chemicals or inhaling fumes.

For some occupations, the potentially toxic chemicals may

be obvious. People who work in chemical or hospital laboratories, photography studios, gas stations or petroleum plants, dry cleaning establishments, or printing plants should already be aware of the strong chemicals and the fumes they contact daily, and they should take serious precautions to protect themselves.

Everyone should look for the less obvious toxins in the workplace, those that appear harmless. The white-collar office is hardly ever considered a source of health hazards. However, there are many, often unrecognized toxins in such settings. For example, a secretary may wish to avoid excessive contact with correction liquids or whiteout products, glue, copy machine powder, or no-carbon copying paper. A commercial artist might choose to be careful with acrylic paints, inks, glue, adhesive sprays, or fumes from computer-set type. Neurologic disease can occur in dentists who deal with amalgam fillings, and the mercury in these fillings can also be toxic to the patient. If you work at a computer on a daily basis, be aware that most of the video display terminals are cathode ray tubes similar to the tubes found in X-ray machines. The computer screen is capable of emitting ionizing X-ray radiation in small amounts. Although this is not a major problem, who needs it? A lead-impregnated mesh screen will offer you protection.

Approach your investigation in a systematic way. Make a list of all the products or objects you work with on a daily basis. Then try to learn the chemical makeup of each to determine if it gives off fumes or will leave a chemical residue on your hands. With many items the risks may be minimal, and you may choose not to alter the way you use them. With others, you may suddenly realize why you suffer from forgetfulness, headaches, or rashes, and you may decide to protect yourself in some manner.

Some very common and potentially harmful things found in offices include yellow lacquered pencils, which may contain lead (avoid them if you are a pencil chewer); fluorescent lighting; the chemicals associated with copying machines; typewriter ribbons, glues, and acetone-based whiteout material; tap water;

and petroleum-based fumes from paint or cleaning fluids (in your office or carried through vents from other offices).

TOXINS IN YOUR NEIGHBORHOOD

Some environmental toxins are beyond the control of individuals and must be dealt with by group effort over time. There is virtually no place in the civilized world that is totally free of environmental pollutants, though some areas do have higher concentrations than others. If you have the freedom to choose where you live, you may be able to move to a safer area than the one you now reside in, but most people cannot do this.

However, we can all educate ourselves about our area and take precautions against toxins in the air, water, and soil. Make a list of industries in your neighborhood or region. What sort of fumes are they emitting into the air? What are they dumping into the soil or permitting to flow into the rivers and streams? Are there nuclear waste dumps or toxic waste sites near your home? If you find too high a number of risks in your area, what steps can you take? Can you join with neighbors to try to improve the situation by pressuring industries and civil authorities? If necessary, can you afford to move?

Statistics indicate that certain areas of the United States have higher rates of death from the man-made diseases than others. It seems clear from Table 9.1 that the high-risk areas for cancer are the more heavily industrialized areas of the northeast and Washington, D.C. Table 9.2 shows a similar pattern for coronary heart disease.

TABLE 9.1
Cancer Deaths by State (1986)

State	Total Deaths	Death Rate Per 100,000 Population
Washington, D.C.	1,700	309
Rhode Island	2,500	263
Florida	29,800	249
Pennsylvania	28,400	241
Massachusetts	13,300	233

TABLE 9.1 (Continued)
Cancer Deaths by State (1986)

State	Total Deaths	Death Rate Per 100,000 Population
New Jersey	17,000	228
New York	38,400	228
Delaware	1,400	227
Connecticut	7,000	224
Missouri	11,100	221
Maine	2,600	218
Ohio	23,200	215
Arkansas	5,200	211
Maryland	9,000	206
Nebraska	3,300	205
West Virginia	4,100	205
Illinois	23,500	205
Alabama	8,300	203
Iowa	6,000	203
South Dakota	1,400	201
Vermont	1,100	200
Kentucky	7,800	199
Tennessee	9,600	196
Indiana	11,000	196
North Dakota	1,300	194
Kansas	4,700	194
Wisconsin	9,500	193
Oklahoma	6,400	193
New Hampshire	2,000	189
Michigan	17,600	188
Virginia	10,600	185
North Carolina	11,500	184
Mississippi	4,800	180
California	47,000	180
Minnesota	7,600	179
Montana	1,500	176
Louisiana	8,000	176
Oregon	5,300	173
Georgia	10,400	169
Washington	7,900	169

(*continued*)

TABLE 9.1 (Continued)
Cancer Deaths by State (1986)

State	Total Deaths	Death Rate Per 100,000 Population
South Carolina	5,700	168
Arizona	5,800	165
Nevada	1,700	155
Texas	24,400	150
New Mexico	2,000	138
Idaho	1,500	135
Hawaii	1,400	130
Colorado	4,300	125
Wyoming	700	114
Puerto Rico	3,000	94
Utah	1,700	88
Alaska	400	84

TABLE 9.2
Coronary Heart Disease Deaths by State (1983)

State	Total Deaths	Death Rate Per 100,000 Population
Pennsylvania	49,911	420
New York	73,868	418
West Virginia	7,880	402
Florida	42,782	398
Rhode Island	3,739	391
South Dakota	2,725	390
Iowa	11,041	380
New Jersey	28,218	378
Missouri	18,538	374
Arkansas	8,617	371
Massachusetts	21,234	369
Maine	4,178	365
Illinois	41,849	365
Ohio	38,763	361
Kentucky	13,178	355
Washington, D.C.	2,210	355

TABLE 9.2 (Continued)
Coronary Heart Disease Deaths by State (1983)

State	Total Deaths	Death Rate Per 100,000 Population
Nebraska	5,267	353
Wisconsin	16,411	346
Kansas	8,332	343
Oklahoma	11,113	336
Michigan	30,427	336
Indiana	18,201	333
Connecticut	10,438	333
Tennessee	15,313	328
Mississippi	8,426	326
Delaware	1,967	325
Vermont	1,703	324
North Dakota	2,159	317
Alabama	12,447	314
New Hampshire	2,994	313
North Carolina	18,817	310
Maryland	13,312	306
Minnesota	12,555	303
Oregon	7,986	301
Louisiana	13,266	299
South Carolina	9,673	297
Virginia	16,156	291
Georgia	16,573	289
Montana	2,245	276
Washington	11,687	272
California	66,752	265
Arizona	7,655	258
Texas	39,973	253
Idaho	2,497	253
Nevada	2,089	233
Colorado	6,430	204
Wyoming	1,001	194
New Mexico	2,588	185
Utah	2,940	182
Hawaii	1,758	173
Alaska	371	77

Guidelines for the Use of Toxins
- Read the warnings on products before you buy, and compare labels of similar products.
- Select the least hazardous product if you have a choice. Water-base paint is less toxic than oil-base paint.
- You do not need a different cleaner for every kind of cleaning problem. One cleaner can serve many purposes.
- Never mix different products. Bleach mixed with ammonia or acids can release chlorine gas.
- Buy only what you need and use up the product completely.
- Recycle what you can. Used automotive fluids should be taken to a recycling center.
- Do not put toxins, including unused paint, pesticides, solvents, and leftover chemicals, in the trash. They will contaminate "sanitary" landfills. Have them collected by your sanitation department for disposal.
- Do not dispose of toxic material by pouring down a drain, on the ground, or in the street. Wastes will contaminate the land, groundwater, lakes, and oceans.

WHAT YOU HAVE ACHIEVED THIS WEEK

This week's efforts cannot be effectively measured, for they have been primarily a matter of education. If you have increased your knowledge of toxins in your environment, you have taken a major step toward minimizing your contact with them. From now on, each time you use something with toxic potential, you should be aware of it and find ways to take precautions.

Chapter 10

Week Seven: Prescription and Nonprescription Drugs

Projects:
1. Inspect your medicine cabinet.
2. Make a list of the drugs you are taking.
3. Begin to eliminate all over-the-counter drugs used on a regular basis.
4. Talk to your doctor about cutting back on your prescription drugs as your health improves.
5. Investigate nonchemical alternatives to some of your perceived drug needs.

If you regularly take any kind of drug, whether aspirin or allergy medication or tranquilizers, you are ingesting chemicals that may have long-range toxic effects. Furthermore, many pills don't even solve the problem you have. For some people, certain drugs may be essential, but for most they are not. No drug can be taken safely without potential side effects or dangers, particularly if it is used over a long period. This week you are to begin eliminating over-the-counter medications and any prescription drugs you and your doctor agree you can do without.

NONPRESCRIPTION DRUGS

In the twentieth century, we have grown accustomed to taking a pill or swallowing a spoonful of liquid or rubbing ourselves with ointment for the slightest discomfort. However, aches, pains, itches, and sneezes are our bodies' ways of telling us there are problems. By taking drugs, we usually get rid of the symptoms rather than the conditions underlying them. Unless the symptoms come from a cold or other infection, these conditions may often be related to the toxins in our environment and diet, and we can eliminate the need for drugs by altering our lifestyle.

Many of the nonprescription drugs we use—and abuse—are made from chemicals that are potentially toxic. Frequently these drugs are worthless, ineffective, or dangerous. Some of them contain multiple chemicals—a sort of shotgun approach in which synergistic action may increase the risk of allergies or other reactions without increasing effectiveness. According to Joel Kaufman and the Public Citizen Health Research Group, even some of the top 40 sellers—including such famous names as Anacin, Listerine, Nyquil, Preparation H, Excedrin, Dristan, Scope mouthwash, Robitussin cough medicines, and Sinutab—contain ingredients that are either unsafe or ineffective. It is extremely important to be aware of this and keep the use of such drugs to a minimum—if they are used at all. The average American family spends in excess of $40 a year per person for these toxic, sometimes worthless drugs. Since all we have to do to obtain these toxins is to walk into a drugstore and purchase them, we must take full responsibility for their use.

Pain Relievers, Fever Reducers, Anti-inflammatory Drugs

Aspirin (acetylsalicylic acid, or ASA) traces its origins to Hippocrates, who discovered 2,400 years ago that the bark of the willow tree had some medicinal value as a diuretic. Almost 2,000 years later, in 1763, the Reverend Edward Stone of England found that the bark could also lower fevers and ease aches and pains. The active ingredient in the bark was found to be salicin, and in 1853 German chemist C. von Gerhardt first synthesized salicylic acid, a derivative of salicin, from phenol. He also prepared acetylsalicylic acid (later named aspirin) and hundreds of other compounds, but with no particular use for them, they gathered dust on the Bayer Company's shelves. It was not until 1899 that Felix Hoffman, also associated with Bayer, rediscovered acetylsalicylic acid and used it as a medicine to treat his ailing father.

Aspirin has generally been considered one of the safest, most effective drugs available. It has been used as a pain reliever, fever reducer, and anti-inflammatory, and as a "blood thinner" to prevent heart attacks. However, aspirin can cause serious reactions, including rashes, asthma, shock, stomach irritation, and bleeding. It is unsafe for children with influenza and chicken pox because it has been linked to the subsequent development of Reye's syndrome, a frequently fatal disease. Aspirin is unsafe in combination with many other drugs, among them anticoagulants (blood thinners), cortisone, gout medication, oral diabetes medication, sulfanilamide, methotrexate, and alcohol.

Aspirin is often sold in combination with other drugs such as antihistamines or caffeine. These combinations are not safe and are not recommended. Various buffered aspirin products such as Ascriptin and Arthritis Pain Formula contain aluminum and should be avoided.

Acetaminophen is the active ingredient in most nonaspirin pain relievers. Like aspirin, acetaminophen is a fever reducer, but it does not relieve inflammation. For this reason it is less effective in treating arthritis. Some common brand names of acetaminophen are Datril, Tylenol, and Panadol. It is used mostly to avoid the problems associated with aspirin, but even this drug is not entirely safe; overdoses can cause liver damage.

Phenacetin, another pain reliever that has been on the market for many years, has always been unsafe; it can cause kidney damage and cancer. The FDA has only recently ordered it removed from the market, but it is still available outside the United States.

Antihistamines, Allergy Medications, Cold Remedies

Colds are respiratory illnesses caused by a virus. They should be treated with lots of fluids—distilled water or fresh fruit juice— and rest. Most cold medicines do little to reduce the average length of the illness (a week), and some are harmful. They do little more than relieve stuffy noses, or aches and pains.

Allergies manifest themselves in sneezing, wheezing, rashes, headaches, and other symptoms. They are best treated by avoidance of contact with the causative factor—including various foods as well as the chemicals in food and in the environment. There are a few nonprescription antihistamines such as chlorpheniramine and brompheniramine that are symptom relievers and may be taken for a few days but no longer.

The usual cough medicine, cold remedy, or allergy medication contains a combination of antihistamines, aspirin or acetaminophen, caffeine, dextromethorphen, and alcohol (if it is a liquid). These multichemical drugs are very popular but not very effective. They should be avoided. Antihistamines do not work for colds and should be avoided. Asthma is a disease that should be treated by your doctor, not with over-the-counter drugs.

Weight-Loss Formulas

Phenylpropanolamine and benzocaine are the main ingredients in most over-the-counter weight-loss products. These are hazardous chemicals that can cause high blood pressure and kidney disease and should be avoided, as should low-calorie foods with synthetic sweeteners.

Constipation Remedies

Constipation is a decrease in the normal number of daily bowel movements or difficulty in passing hard stools. Medical attention is suggested if bowel habits suddenly change or if blood or black stools are passed. Self-treatment of constipation with laxatives is not recommended. Usually a low-toxin diet (or any high-fiber, low-fat diet) will solve constipation problems within one week, without resort to drugs. With a high fiber intake, stools have a transit time of approximately 24 hours and will pass easily.

Certain drugs cause constipation, including narcotics and cough medicines with codeine, cholestyramine and other drugs to lower cholesterol, iron supplements, tranquilizers, sleeping pills, antihistamines, and water pills. This week is a good time to be aware of them and plan to discontinue their use. Bulk-forming laxatives such as Metamucil, psyllium, or bran may be used for short periods of time but are usually unnecessary if your diet is high in fiber.

Nausea, Vomiting, and Diarrhea Remedies

Simple nausea and vomiting can be a reaction to food, motion, drugs, or alcohol, or to an infection transmitted through tainted food or water. Persistent nausea or vomiting may occur with pregnancy or many diseases. These require treatment by your doctor. Diarrhea is usually a temporary discomfort caused by intestinal infection, allergic reaction, chemicals in food, or some medications.

Ordinarily nausea, vomiting, and diarrhea clear by themselves in one to three days and require no treatment other than fluid and body salt replacement. Drugs are not usually necessary. However, any symptoms that do not clear promptly should be checked by your doctor.

Antacids are frequently used for "indigestion," stomach upset, and ulcers. Many antacids, including Maalox, Gelusil, Amphojel, Mylanta, Rolaids, Di-Gel, and various others, contain aluminum hydroxide or other forms of aluminum. Since there is a possible association between aluminum and Alzheimer's disease, antacids of this this type should be avoided. Titralac and Tums are among the antacids that contain calcium but no aluminum.

Sleep Regulators

The bill for nonprescription sleep aids in 1981 for all Americans was over $30 million, so the problem of insomnia appears to be a big one. However, these drugs are of very little use. In most instances insomnia can be handled without drugs. Such simple measures as exercising, eliminating alcohol and stimulants, and eating simple meals earlier in the evening will solve most minor insomnia problems.

PRESCRIPTION DRUGS

The fact that a prescription is required to obtain a particular drug indicates that there is some risk involved in its use. In prescribing a drug, the doctor generally weighs its advantages against its risks, but unless the patient asks, the doctor will not usually explain the relative values and dangers of a medication. Sometimes the doctor may not even be aware of all the potential side effects of all the available drugs—particularly when they are taken along with other drugs.

In some cases, a drug's side effects are not known to anyone until considerable harm has been done. An example is thalidomide, which was introduced in European countries in 1961. Used as a mild sedative during early pregnancy, thalidomide resulted in catastrophic birth defects.

And most medications do have "side effects," whether simply mood changes or "allergic reactions" such as rashes or more serious symptoms. Allergic reactions to drugs are very common, but they represent only 25% of all the side effects associated with drugs, which can range from anemia to neurological damage. Nearly 40% of people taking prescription medication outside hospitals suffer from side effects. Before the proliferation of drugs, an estimated 5% of all hospital admissions were a result of drug reactions; in recent years, this figure has increased to approximately 20%. Approximately 30% of all hospitalized patients develop drug reactions, including those who have been admitted for drug reactions in the first place.

You should become aware of the side effects of the drugs you take, so that you can participate in deciding which drugs are truly of benefit. Your doctor's advice should continue to be important, especially regarding serious diseases or illnesses.

Sometimes a drug's side effects can be more serious than the condition it is designed to treat. The risks of an illness have to far outweigh the risks of the drug itself before the drug should be prescribed or taken. Many drugs, if not all, are xenobiotics and are therefore potentially toxic; consequently they may have long-range deleterious effects such as the development of cancers, Parkinson's disease, lupuslike diseases, autoimmune disease, aplastic anemia, earlier heart attacks, etc.

Some drugs are known or suspected to cause cancer in laboratory animals and humans. These include diphenylhydantoin (used in the treatment of epilepsy); the hormone diethylstilbestrol (DES), oral contraceptives, and "body-building" anabolic steroids; amphetamines; the antibiotic chloramphenicol; the analgesic phenacetin; methoxypsoralen (used with ultraviolet light to treat psoriasis); cholestyramine (used to lower cholesterol); and such hydrazine derivatives[1] as nitrofurantoin, phenelzine, hydralazine, and iproniazide (some hydrazine drugs may also cause lupus, birth defects, and hemolytic anemia). Even drugs used to treat cancer may ultimately cause additional cancers.

Since drugs can be responsible for all sorts of symptoms, it frequently becomes difficult to separate reactions to medication from the disease being treated. The simultaneous use of many pills and capsules may further muddle cause and effect.

If you are currently taking prescription drugs, make a list of

---

[1]In addition to their pharmaceutical applications, hydrazine derivatives are used in plastics, rubber products, anticorrosives, herbicides, pesticides, photographic supplies, preservatives, textiles, dyes (Tartrazine Yellow #5, among others), and rocket fuel. Hydrazine is found naturally in a few plants, including mushrooms and tobacco.

them. Ask your physician what specifically they are supposed to do for you and whether there is some alternative to their use. Also, attempt to learn what side effects these drugs may have.

I am not suggesting that you discontinue all medications. Whether to take or discontinue medications should be decided between you and your physician. However, where possible, it is highly appropriate to use a dietary approach to *prevent* the major noninfectious illnesses. When you are being treated with drugs for such illnesses as gout, high blood pressure, coronary heart disease, diabetes, and others, your need for the drugs will decrease as your health improves through diet alone.

Remember that certain nondrug testing or treatments may also offer risks that outweight the potential benefits. Certainly, X-rays and radioactive isotopes are important in the diagnosis and treatment of serious illnesses such as cancer or leukemia. But X-rays and radiation should not be used lightly for "routine" examinations or minor ailments. X-radiation and isotope radioactivity create free radicals within the body; these may cause long-range damage to tissue, DNA, and RNA, and may result in the development of cancers, immune system damage, and aplastic anemia, among other problems. In the past, X-rays were used on patients far more frequently than they should have been.

ALTERNATIVES TO DRUGS

Many minor and some major health problems can be at least partially resolved without medication. Dietary adjustments can frequently have more miraculous effects than any "miracle" drug. The day-to-day problems of obesity, constipation, rashes, insomnia, and headaches can be resolved almost entirely by diet.

After beginning a low-toxin diet in combination with an exercise program, many diabetics find it necessary to cut down on oral hypoglycemic agents or insulin dosage. And people with high blood pressure find they are able to control it without drugs once they restrict animal foods and salt intake and lose weight through proper diet and exercise. Dietary measures can also be used to treat gout without drugs.

Some people have found that exercise, attitude, relaxation techniques, and behavior modification programs result in improved health and thus reduce dependency on drugs and medications. In many cases biofeedback training has been particularly helpful in controlling high blood pressure and asthma, among other ailments.

WHAT YOU HAVE ACHIEVED THIS WEEK

This is one of those weeks when it will be difficult to measure your accomplishments. If you have managed to discontinue use of any drugs—prescription or nonprescription—simply be aware that you have reduced your intake of toxins and you will actually feel better. The avoidance of drugs, combined with your reduced intake of other poisons, will have a cumulative beneficial effect.

Chapter **11**

Week Eight: Eliminating Seafood and Poultry

Projects:
1. Stop eating fish, seafood, and poultry.
2. Stop buying all other meats.

Fish, seafood, poultry, and all other animal foods are on the high end of the food chain. Because of this they tend to concentrate xenobiotics from the environment. To remind yourself of the level of toxins found in various foods, review Table 1.2.

This week you will begin removing meat from your diet. This will be the most difficult part of the program for many people, and for this reason two to three weeks are allowed for the gradual elimination of all meats, poultry, and fish, beginning this week with poultry and fish.

This week you should use up whatever poultry and fish you have in stock and remove them from your shopping list. You should also stop ordering fish and chicken dishes when you are dining out.

It is important to get rid of all forms of fish, not just fresh or frozen fish served as entrées. Canned fish (tuna, sardines, salmon, etc.) used for salads or sandwiches should be removed from the diet, as should such breakfast fish as lox and herring, and fish in soups, chowders, and gumbos. This also applies to shellfish—any kind of seafood.

As far as poultry dishes are concerned, all should be discontinued, including chicken soup, barbecued chicken, chicken livers, and skinned white meat of chicken. Even the annual Thanksgiving turkey should be avoided by those who are chronically ill; if you are healthy, celebrating with a traditional meal several times a year will not create any major problems as long as you have remained on the program otherwise. However, before you purchase that holiday bird, be aware that a recent survey showed that 36% to 58% of poultry processed, inspected, and sold in the United States is contaminated by salmonella organisms, which can cause serious intestinal disease.

Fish caught in waters close to industrialized areas are often contaminated by heavy metals, pesticides, industrial discharges, and treated effluent from municipal sewage disposal. Shellfish off the coast of California have been monitored for contamination of various sorts in recent years. They have been found to have high levels of lead, silver, zinc, PCBs, and pesticides, as well as a chronic low level of oil contamination. Fish caught mid-ocean, away from shores where industrial wastes such as DDT and PCBs have been discarded, *may* be free of these xenobiotics. However, fish also store large amounts of the methylmercury found naturally in ocean waters. Large fish such as blue marlin and swordfish have the highest levels of methylmercury; tuna, shark, red snapper, and grouper are intermediate; pollock, salmon, bass, sole, flounder, and shellfish have lower levels.

The methylmercury problem is not limited to ocean fish. Acid rain, which affects large areas of the world today, is capable of causing the release of mercury from rock formations and soil. This mercury washes into lakes and streams where bacteria can convert it into methylmercury. Through the gradual movement of methylmercury up the food chain, freshwater fish will ultimately contain high levels of methylmercury. Consequently, the Swedish government has seen fit to ban fishing in many Swedish lakes. High levels of mercury contamination were found in fish caught in the Great Lakes area during the 1970s; fish caught in apparently pristine lakes high in the Rocky Mountains also contain methylmercury.

Recent research indicates that silver-mercury amalgam dental fillings release absorbable mercury vapors into the body. Since the toxic effect of mercury from all sources is cumulative, if you have amalgam fillings (and almost all of us do), be particularly careful to avoid seafood or fish. Be sure to follow my recommendations for taking selenium, ascorbic acid, and vitamin E, for they protect against many of the toxic effects of mercury.

PCBs, mercury, and other harmful xenobiotics found in fish may be particularly harmful to pregnant women, nursing mothers, and their infants. These toxins can cause birth defects and may also be responsible for sterility. If you are planning to have a child, it is especially important that you stop eating all fish and seafood; if you have always been a fish eater because it was considered "healthy," if you have been exposed to xenobiotics from other chemical sources (pesticides, solvents, tobacco smoke, contaminated tap water, etc.), or if you have developed gestational diabetes, you should consider following the ten-point program in its entirety.

UPDATE YOUR SHOPPING LIST

Get out your pen to make what will be among the last of the changes in your shopping list. Cross out fish and shellfish: shrimp, oysters, clams, scallops, crab, lobster, salmon, herring, tuna, sardines, halibut, swordfish, cod, sole, catfish, trout, etc. Do the same for all poultry and poultry products, including chicken, turkey, Cornish hen, squab, and game fowl.

Add more fresh fruits and vegetables.

Remember, eggs and dairy products should have been eliminated already. Don't forget that ice cream and yogurt should not be on your shopping list.

WHAT YOU HAVE ACHIEVED THIS WEEK

If you have stopped eating all forms of poultry and fish this week, you are no longer getting the hormones, pesticides, DDT, salmonella organisms, PCBs, methylmercury, and other toxins present in fish and poultry.

Chapter 12

Week Nine: Eliminating Red Meat and Fats

Projects:
1. Remove pork, beef, lamb, and most fats from your diet.
2. Complete the transition to a low-toxin diet.

This may be the toughest week for many people. If you are addicted to juicy barbecued steaks, fast-food hamburgers, or Coney Island hotdogs, don't despair. Just be aware that some foods are more difficult to give up than others, and take your time. Allow an extra week or two if necessary. Remember that you will be excluding all forms of beef and pork this week—steak, roast beef, corned beef, veal, tongue, ground beef, pork roast, ham, frankfurters, sausage, and meat salads, soups, and stews.

As of this week, you are almost entirely on the low-toxin diet. If you have been following the regimen so far, you should be feeling much more vigorous and energetic than in the past, and that should give you the incentive to get through the remaining weeks without straying.

FAT AND CHOLESTEROL

In recent years, a great deal of attention has been devoted to fat and cholesterol in the diet. They have been blamed for the increase in coronary artery disease, as well as for several other ailments. Considerable research and experimentation have linked dietary consumption of them with these diseases, yet there is no satisfactory explanation for why fat and cholesterol, so important for sustaining life and bodily functions, should be detrimental to health. Nor does there seem to be a logical reason why milk, meant to feed the young of all mammals, and eggs, the almost perfect source of nutrition, should also be connected to rising levels of human blood cholesterol and ultimately to coronary heart disease.

The search for an explanation must be directed by the fact that these essential nutritional elements have not always caused deadly diseases. The origin, development, and rising incidence of the diseases linked to fat and cholesterol can be traced. Some examples have already been given in Chapter 2, but there are numerous others.

In the South Pacific, coconuts high in saturated fats were always a food staple for the natives, but until "civilization" arrived, they did not cause coronary heart disease. Now saturated fats in coconuts are linked to this condition.

From a medical standpoint, the example of China is particularly significant, for studies of the Chinese diet were influential in the development of some current attitudes toward fat and cholesterol. These studies were among the many important contributions made by my professor from medical school days, Dr. Isidore Snapper. Dr. Snapper was of Dutch origin but spent most of his life practicing medicine and doing research in China before World War II. His work there ended when he was forced to leave the country by the Japanese invasion, and eventually he emigrated to the United States.

Throughout his many years of caring for patients and doing research at a large Chinese hospital, Dr. Snapper had found only one victim of coronary heart disease. He observed that the Chinese diet consisted predominantly of vegetables and rice, with a minimum of red meats. He also noted that the main fat consumed by the Chinese was unsaturated vegetable oil, and he attributed the low rate of coronary heart disease to this fact.

Partly as a result of Dr. Snapper's important observations, in the 1950s industrialized countries began to turn to polyunsaturated corn and safflower oils, hoping to improve national health. However, something did not quite translate. Even though blood cholesterol levels decreased with the use of polyunsaturated oils, the incidence of coronary heart disease did not drop, and cancer death rates and the incidence of gallstones actually increased.

It does not diminish the significance of Dr. Snapper's work to point out that there were factors he failed to take into consideration. First, the environment of the industrialized nations was not the same as that of the agricultural China he knew. There were vast differences in farming methods, the West utilizing industrial technology and China using techniques that were centuries old.

But perhaps more significant was the fact that all Chinese did not adhere strictly to the simple diet observed by Dr. Snapper. The wealthier Chinese did indeed eat red meats such as lamb, beef, and pork, as well as chicken in all forms, with all the fatty tissue. They also ate duck, a high-fat meat, cooked Peking-style, with crispy skin a particular delicacy. All levels of Chinese society enjoyed the eggs of all types of birds—chickens, ducks, geese, pigeons, and seagulls—and the wealthier citizens relished eggs buried in the ground until they turned black and odiferous from sulfur. This is precisely the sort of diet that causes heart attacks, according to modern theories, yet heart attacks were virtually unknown in China.

The only logical conclusion is that it is not fat and cholesterol in themselves that cause health problems but something found within them only in industrialized societies. This something is clearly the chemical toxins created by man and his technology, many of which have already been linked to the killer diseases. We know that many toxic chemicals are fat-soluble and can therefore be easily stored and accumulated in fatty tissue, gradually becoming more and more potent poisons.

But what are these poisons, and how do they get into the animal foods of industrialized societies?

INDUSTRIAL FARMING TECHNIQUES

Before the Industrial Revolution and up to the twentieth century, farm animals were permitted to graze and forage at will, acquiring their food from nature. This natural supply of food and water was largely unpolluted by chemical wastes, though some parts of the world were gradually becoming contaminated.

But as the twentieth century progressed, the food supplies of domesticated animals became increasingly tainted with man-made toxins. Soil and groundwater throughout the industrialized world became polluted with chemical waste from factories, and farmers increasingly used fertilizers and pesticides. Traces of all of these xenobiotics began to appear in grasses and grains grown to feed cattle.

But the tainting of our meat sources of food did not end there. With the twentieth century also came the concept of mass production of food animals—cows, hogs, sheep, and poultry. Manufactured feeds prepared from grain sprayed with insecticides began to be used to supplement grazing, or to replace it entirely in cases where animals were no longer allowed to roam free but were confined to increase their fat content and make them more profitable.

As an additional growth stimulant, animals are also given antibiotics and hormones, which raise levels of contamination in animal tissue and thus in the human diet. The use of hormones has created very specific problems for some people. Eating hormone-treated chicken has caused sex characteristics—including fully developed breasts, pubic hair, and menstrual periods—to appear in girls under the age of two. In some cases, young boys have developed female sex characteristics. The same hormone, diethylstilbestrol, has also been responsible for the development of cancers in the daughters of women who took it during pregnancy.

The chemicals from all these sources remain fairly intact in beef, lamb, pork, and poultry, stored in the animals' organs, muscles, cholesterol, and fat, and are passed along to the consumer. For example, to take just one group of toxins, animal products are the single largest source of pesticides in the human diet; because of bioconcentration and biomagnification, we obtain from meat 16 times the amount of pesticides we would from an equivalent amount of plant food.

As if the abovementioned toxins were not enough, beginning in the 1950s, grease and used fats complicated the picture. Previously, fats used for frying foods—lard, vegetable oil, or shortening—were collected from restaurants and other eating establishments for use in the manufacture of soap. With the development of detergents, these fats were no longer needed by the soap industry. Rather than discard them, it was decided that they could serve as a food supplement for livestock. But the effect of reintroducing them into the food chain is to concentrate the poisons in the fats even further.

The meat from cattle, hogs, and poultry raised in industrial lands is not the only animal food source we must be concerned with. Animals transmit toxins, including PCBs, DDT, and radioactive substances, to their eggs and milk as well. This is true not only for domesticated animals but especially for humans; very high levels of xenobiotics have been reported in human breast milk. Humans are at the top of the food chain and receive toxins from all food sources. It is therefore advisable to avoid all animal sources of food, not just meat and meat products.

FATS AND THE LOW-TOXIN DIET

In the early part of this century, coronary heart disease was rarely seen,[1] yet in 1920 the average daily calorie intake was greater than it is today, the consumption of butter, a primary source of saturated fats, was three times greater, and the consumption of eggs was two times greater. Polyunsaturated fats

[1]Coronary artery occlusion was first brought to the attention of the medical profession in the United States by Dr. James Herrick of Chicago in 1912.

and oils, occasionally recommended today as a means of avoiding coronary heart disease, were used half as frequently then.

Current treatment of and research into coronary heart disease are concerned primarily with either lowering dietary intake of fats and cholesterol or reducing the blood cholesterol level by the use of drugs. Scientists have not seriously considered the possibility that the real culprits are the xenobiotics stored in the fats. It is suggestive, however, that nicotinic acid—a B-complex vitamin and, I believe, an antioxidant—has been used to lower cholesterol levels and has been shown to prolóng life in those who have had heart attacks, perhaps because it may be neutralizing free radicals.

Of the organizations involved in research and fundraising for various diseases, a few offer dietary recommendations, but their advice differs substantially from the low-toxin program. This is not surprising, since most of these organizations are not attuned to the concept of xenobiotics in food as a potential cause of disease.

For example, the American Heart Association recommends a diet that limits the intake of fats to 30% of total daily caloric intake: 10% saturated fats (the ones found in the foods most people crave—ice cream, steaks, chicken, and cheeses), 10% polyunsaturated (from vegetable sources), and 10% monounsaturated (found primarily in olive oil and avocados). The ten-point low-toxin program assumes that it is not fats and cholesterol that have undesirable effects but what has gotten into fat and what has been stored in muscle and other tissue.

This week you should eliminate all animal sources of fat from your diet. As a compromise, you can use small amounts of vegetable oils. Although the monitoring of pesticides in food has indicated that vegetable oils are also carriers of toxins and have higher levels than the vegetables themselves, the amount is considerably less than in animal tissue and dairy and egg products (see Table 1.2).

Polyunsaturated vegetable oils like safflower, sunflower, and corn oil are far safer than animal fats, and they have the additional benefit of containing some essential fatty acids. However, they are not without risk, especially if overused. They react adversely to free radicals, rapidly becoming rancid, and their use has been linked to cancer.

Polyunsaturated vegetable oils and monounsaturated olive oil should be used infrequently and in small amounts, as occasional minor additions to salads and ethnic dishes where they are needed for special flavors. Cottonseed oil, however, should never be used for food purposes because more pesticides tend to be sprayed on a nonfood commodity like cotton. Moreover, all vegetable oil sources are not alike. Cocoa butter, palm oil, and coconut oil are very highly saturated and in our era have been linked to elevated cholesterol levels.

Polyunsaturated fatty acids, including the omega-3 group—eicosapentaenoic acid (EPA) and docosahexaenoic acid (DHA)—are also obtained from cold-water fish such as salmon. Research has shown that the polyunsaturated fatty acids found in fish oils are capable of decreasing the death rate from coronary heart disease, and this is the supposed reason why Eskimos are protected. Fish and the omega-3 group of supplements may be beneficial—if they contain no PCBs, DDT, mercury, or other xenobiotics. The studies, however, have not shown whether the fish eaters succumb to other diseases such as cancer instead of heart attacks. Thus, we should reserve judgment on fish oils until we are certain about their efficacy and safety.

Hydrogenated or partially hydrogenated oils of all types should be avoided. Hydrogenation forces hydrogen atoms under high heat and pressure to fill the unsaturated links in vegetable fat molecules, thus converting a liquid fat into a solid or a semisolid at room temperature, and creating an artificial saturated fat rarely found in nature. Because of free-radical oxidation, unsaturated fat turns rancid if left too long on supermarket shelves, and this presents a problem for the food industry. The commercial practice of hydrogenating polyunsaturated vegetable oils in effect does away with this problem, but in the process it destroys all the essential fatty acids.

The process for the hydrogenation of fats won a Nobel Prize in 1912 for French chemist Paul Sabatier, who developed it to make a low-cost soap from waste fish and vegetable oils instead of expensive lard. It was American chemists who adapted the process for food purposes such as the manufacture of margarines and shortening. Hydrogenated oils are found in many processed foods, including baked goods and breakfast cereals, ice cream, chocolate candies, cakes, chips, and breads. However, these fats are not easily converted to usable fats by our body enzyme systems.

UPDATE YOUR SHOPPING LIST

Finish using all meats you have in stock this week, and remove them from your shopping list. Many of these should have been removed some time ago—breakfast steaks, breakfast ham, bacon, sausage, processed meats, and canned or frozen meats. If these were not removed earlier, do so now, and cross off the other meats as well—lamb, mutton, steaks, roast beef, veal, ground beef, pork roast, and ham.

WHAT YOU HAVE ACHIEVED THIS WEEK

If you have completely eliminated beef, pork, and lamb from your diet this week, you have made a major accomplishment. The quantity and kind of toxins you have removed from your diet cannot be accurately measured, but rest assured that they are numerous. You are also no longer ingesting the antibiotics and hormones injected into cattle by the raisers. And, no small achievement in itself, you are reducing your blood cholesterol level.

This is the final step in making the transition to a diet composed only of plant sources of food. By now your taste buds should be finding that vegetables have a satisfying flavor, and you should be feeling much healthier and more energetic.

Congratulations! You are well on your way to good health!

Chapter 13

Week Ten: Dining Out

Projects:
1. Eat breakfast in a restaurant.
2. Eat lunch in a restaurant.
3. Eat dinner in a restaurant.
4. Have a dinner party.

At last you've made it to the end of the program. You have learned how to minimize your consumption of chemical toxins, which should increase your chances of living a longer and healthier life than you otherwise would have. The final challenge is to stay on the low-toxin diet when you go out to restaurants. Plan to dine out at least three times this week—for breakfast, for lunch, and for dinner. At the end of the week, give a big dinner party to test your ability to entertain your friends while serving healthy foods only.

Dining out in commercial establishments has been a custom for many centuries. There is a record of a restaurant operating in ancient Egypt that offered a single choice to its patrons—a plate of cereal, wildfowl, and onions. For travelers, there have been taverns and inns in the European nations since the Middle

Ages, and fine restaurants became a highlight of cosmopolitan cities in the nineteenth century. The first American restaurant in which patrons sat at tables, were served by waiters, and could select meals from an extensive menu was Delmonico's, which opened in New York in 1827.

However, it was not until World War II and after that eating out in restaurants became a truly widespread habit involving all classes of people on a regular basis. During this period, family life was disrupted, and dining out gained wide acceptance.

Today, at least one out of every four dollars of the American food budget is spent in restaurants. Though it is cheaper to eat at home, cost is a minor consideration, for dining out is a means of socializing with friends or family. Women are pursuing new careers away from home, and in many families it is not unusual for both spouses to work. Cooking has become a chore not relished by working wives, so home-cooked meals are less common than in the past.

Dining out has now become a necessity; there is virtually no way of avoiding it, even if we want to. But this presents a special problem for people who want to stay on a low-toxin diet, for a restaurant can be a virtual minefield of xenobiotic chemicals. I have therefore developed special guidelines to help you make choices when eating breakfast, lunch, and dinner in restaurants.

SELECTING A RESTAURANT

You can generally find something to eat in almost any kind of restaurant, even if only a salad. Even some fast-food restaurants now have salad bars. However, your dining out will be more pleasurable if you choose restaurants that have a selection of low-toxin items on the menu. If you are in doubt, call the restaurant one or two days in advance and tell them that you are on a restricted diet. Very frequently they will be happy to make adjustments for you. The same goes for banquets or meetings; by calling the caterer, hotel, or restaurant in advance, you can generally assure yourself of obtaining a low-toxin, fat-free meal.

Vegetarian and Health Food Restaurants

If you are lucky enough to be in a city that has vegetarian or heath food restaurants, you will find that they offer numerous items to choose from. However, many vegetarian restaurants serve foods prepared with high-toxin ingredients like eggs, cheese, butter, and yogurt. They may also fry various items with cooking oils. Determine which foods fit these categories and avoid them or ask the chef to omit them for you. You should not be embarrassed. Most restaurant owners are happy to provide you with a satisfying meal.

Ethnic Restaurants

Many of the ethnic restaurants, especially Indian, Italian, French, and Chinese, are excellent possibilities for low-toxin meals as long as you determine how the food is prepared before ordering. Eggless pastas made from unbleached durum wheat flour are frequently available. Meatless tomato sauces prepared with vegetables or mushrooms are standard in many Italian restaurants; the sauces can also be prepared with little or no oil. Ask the owners to add such sauces to their menus. Minestrone, lentil, and bean-and-macaroni soups are also commonly available. Steamed or stewed vegetables and baked potatoes can be cooked to order in most French restaurants, and you can always have steamed rice or noodles in Oriental restaurants. Since many Chinese restaurants prepare their food to order, you can enjoy exotic dishes of Chinese vegetables steamed without the addition of chicken broth, monosodium glutamate, or soy sauce containing sodium benzoate preservatives. If you *must* have your vegetables stir-fried, be sure that your Chinese restaurant uses small amounts of peanut oil and not lard (unfortunately, the use of lard is common). You may even wish to bring your own preservative-free tamari soy sauce.

Mexican restaurants, Jewish delicatessens, and Greek and

Middle Eastern restaurants may pose greater problems. However, salads, fruits, and vegetable dishes are still possibilities.

Steak and Seafood Restaurants

Among the most difficult places to find a wide selection of low-toxin dishes are steak and seafood restaurants, as well as those that feature roast beef, barbecue and ribs, and hamburgers. However, you may find yourself in situations where you cannot avoid these restaurants, so look for baked potatoes, salads, and vegetable dishes.

Airline Food

If you are a frequent flier or travel occasionally by air, it pays to do some advance planning. Order your meal at least 24 hours in advance. Most airlines routinely prepare special meals for their patrons, whether for religious or health purposes. However, ordering a vegetarian or low-fat meal is no guarantee since the commissary preparing the food may not always understand exactly what you want. If you have had difficulty in the past, order a fresh fruit plate. There is little room for misunderstanding, and the meal is often quite satisfactory.

ORDERING MEALS

Breakfast

Of the three daily meals, breakfast is usually the easiest to order in a restaurant. Although many restaurants feature the standard high-toxin breakfast foods—eggs, bacon, sausage, fried hash-browned potatoes (usually heavily sulfited), coffee, synthetic cream substitutes, and heavily sugared cereals—there are usually a few good items on the menu. Many places serve freshly squeezed fruit juices. You can sometimes get whole wheat pancakes, but do not use butter with them. Try some applesauce instead. Bring your own if you know that it is not available at the restaurant. Toast made from whole wheat or sourdough bread is another choice, as is oatmeal or wheat cereal. Fresh fruit is almost always available.

Lunch and Dinner

When it comes to lunch and dinner, you may find it necessary to speak to your waiter or waitress if it seems there is nothing for you on the menu. Most salads are acceptable, but have them omit any meats or cheeses. You may wish to eat your salad with a simple lemon and herb dressing. Lentil, split pea, and vegetable soups are good choices if there is no beef or chicken broth in the stock.

Among the restaurant breads that may be acceptable are sourdough, crusty French or Italian, some whole wheat breads, pita, and corn tortillas made without bleached flour or preservatives. However, avoid white bread or any bread made with bleached flour.

Your choice of main course may be somewhat easier. Possible selections include pasta with a vegetable, mushroom, tomato, or marinara sauce; steamed vegetables; baked potatoes; and rice and beans. There are many interesting vegetarian dishes available in Indian restaurants. Whatever dish you order, it should be prepared without cheese, bleached sugar, or fats.

Most desserts, including puddings, ice cream, and creamed pastries, are unacceptable on the low-toxin diet. However, some restaurants offer fresh fruit, which is of course the ideal selection.

Guidelines for Dining Out

- Find out if the water served is tap water (it probably is). If distilled or spring water is not available, order mineral water. Fresh lemon juice, a source of ascorbic acid, may be added to tap water if all else fails.
- Try to obtain freshly squeezed fruit juice as a beverage.
- If you have a choice, order bread made with whole grains or unbleached flour.
- If you are sensitive to sulfites, be aware that they are still

found in a variety of retaurant foods even though they have been banned from use on fruits and vegetables in salad bars. You can test your food for sulfites using the commercially available strips described in Chapter 8.

- Avoid most commercial salad dressings. If you cannot determine how the dressing is made, ask for lemon.
- Request that monosodium glutamate be omitted from your order. If it cannot be omitted, order something else.
- See if a low-toxin dish can be prepared without oils, eggs, cheese, butter, or milk. If it meets all these requirements, you may wish to order it.

PLANNING A LOW-TOXIN DINNER PARTY

Conclude the tenth week of your program by hosting a dinner party featuring low-toxin foods. If you are still somewhat self-conscious about the low-toxin diet, you may wish to invite only close friends with whom you have shared your venture, or people who are willing to try something out of the ordinary.

Plan your menu to include your favorite new dishes (see Chapter 15 for recipe ideas), preparing a full five-course meal so as to offer your guests a wide variety of flavors. Numerous dishes are especially helpful if your friends have not yet tried any of your low-toxin foods.

This dinner party is a time for celebration, for you have now completed a very difficult ten-week program that is setting you on the road to good health and long life.

WHAT YOU HAVE ACHIEVED THIS WEEK

If you have managed to eat three different meals in restaurants and have hosted a dinner party for your friends, in all cases maintaining your low-toxin diet, you have passed an important psychological hurdle. You now know that it is possible to avoid the major toxins in virtually any circumstances. You now realize that you can make the necessary health choices without fear or apprehension.

Chapter 14

Week Eleven and Beyond: A New Way of Life

You are now on your own. You should maintain the low-toxin program for the rest of your life. By this time you should want to do so, because you should feel much healthier and more energetic as the toxin levels in your body decrease. If you were overweight, you should have lost some weight already, and there should be further weight loss to come.

Remember, it is up to you whether you go off the program or stay on it; it is your choice to avoid or consume the poisons in the typical American diet. Guilt is not the Eleventh Commandment, ordained by the low-toxin program. As long as you are aware of the risks involved, you will probably not cheat too often, because you know that the only person you are cheating is yourself.

It is probably impossible for anyone to live up to this strict regimen 100% of the time. If you are in danger of dehydration and no distilled water is available, one or two glasses of tap water are better than no water at all. If, once every three or four weeks, you can't resist having a barbecued steak, that won't suddenly give you cancer in and of itself. Nor will an occasional canapé or hors d'oeuvre with a rare glass of wine or beer at a cocktail party do you in. It is the cumulative effect of all toxins from all sources that concerns us.

If you can adhere to the low-toxin plan 90% of the time, your health will benefit greatly. The goal is to keep your weight, cholesterol level, blood pressure, and exercise schedule within accepted ranges. It is also extremely important to avoid tobacco products and tap water, to take antioxidant vitamins and mineral supplements regularly, and to get an adequate supply of high-fiber complex carbohydrates. And these parts of the program are relatively easy to follow.

Those who cannot avoid environmental toxins, or who cannot give up processed goodies, beef, poultry, fish, eggs, and dairy products, will derive at least some benefit from following the rest of the program while trying to minimize their deviation from the more difficult points. In fact, some benefit will result from following any of the ten points of the program consistently. This is not an all-or-nothing program, but the more points you follow on a regular basis, the more benefits you gain.

For those who are chronically ill when starting the program, it is advisable to be somewhat more restrictive in straying from the recommendations. Not until your health returns should you experiment with the occasional deviation.

And chances are that your health *will* return. Consider the following examples.

Case Study: Professional engineer
Name: Irwin Baker
Age: 58
Problems: Overweight, near fatal arrhythmia
  unresponsive to standard medical treatment

Irwin is a good friend and fellow runner who suffered ten years ago from a recurring heart irregularity. Although he enjoyed running, it was an extreme effort for him to run over half a mile. He had been hospitalized on two occasions, and his most recent episode of arrhythmia had been so severe that he was fully convinced he would not survive an additional attack.

At about this time, he was introduced to the Pritikin diet. Since he was desperate, he was willing to try anything. He asked his cardiologist for permission, and the doctor replied, "Irwin, there is no diet in the world that can help your condition, so I don't want you to have false hopes. But certainly it can't harm you, so go ahead and try it."

Within one month Irwin lost weight, was able to discontinue his medications, and began running again, for his irregular heartbeat was now a thing of the past. He has become a marathoner and has completed a number of ultramarathons.

Irwin began to learn about the low-toxin program during many of our long runs together. This is what he has to say about it: "As an engineer I deal with numbers and facts. I know that the Pritikin program worked for me, but I had a gnawing uneasiness about it because it did not quite come to grips with those societies in the world that do the opposite of what he recommends and still stay healthy, namely the Eskimo and the Masai. What you are proposing makes such good sense that I wonder why no one has thought of it earlier. I now know why your program and the similar programs work. I have modified my Pritikin diet to conform to the low-toxin program by including distilled water, megavitamins, and almost no animal foods."

Case Study: Retired factory worker
Name: Juanita Gomez
Age: 50
Problem: Severe rheumatoid arthritis

"For the last sixteen years I have suffered with pain and swelling in just about all my muscles and joints. At first it occurred on and off, and only in my arms and hands. My regular doctor gave up when it got worse. He sent me to a specialist who did all sorts of tests that didn't mean anything to me except that they were very expensive. I suppose I should be thankful that he diagnosed it as rheumatoid arthritis. However, that didn't help me get well.

"I guess the disease was really serious, because he started treating me with many types of medicines, including gold shots, large doses of cortisone, Anaprox, Feldene, laxatives for the constipation, and so many other drugs that I couldn't keep track of them. None of them seemed to make much difference to my health since my joints continued to get worse. I began to gain weight. My face swelled and became round and puffy from the cortisone, and I didn't look very attractive. I was always tired. Pretty soon I had difficulty walking without the use of two canes. I had to apply for disability since I could no longer work. My hands and wrists became contorted so badly that I had to have surgery in my right hand and wrist because I could no longer use them. The next surprise my doctor had in store for me was a recommendation to have my knee joint replaced with an artificial one. Boy, he really scared me with that. I really didn't want the surgery, but the doctor told me that I had no alternative."

Juanita came to see me for an allergic problem, not her arthritis. Understandably, she was skeptical when I suggested dietary changes to treat her arthritis; there are very few good research papers confirming the value of a low-toxin or vegetarian diet for arthritis. Nevertheless, Juanita was willing to experiment and modify her diet in the hope that it would make her feel better.

Within several months of starting the low-toxin program, Juanita's overall condition had improved dramatically. We gradually reduced her medication. Her weight decreased by 20 pounds, and her attitude as well as her joint mobility began to improve. She decided that surgery on her knee was not needed after all. She was ecstatic and is now a firm believer in avoiding xenobiotics.

One question she asked me frequently: "Why don't doctors use a simple dietary change instead of medicines to help arthritis victims?" The answer is not so simple. I believe that physicians will recognize the importance of diet when they have "proof" from double-blind studies. However, as a medical consumer you may have to take it upon yourself to seek the services of a physician open-minded enough to work with you before such studies become available.

Case Study: Retired engineer
Name: Albert Johnson
Age: 70
Problems: High cholesterol, mild diabetes mellitus, overweight, coronary heart disease, high blood pressure, impotence, Ménière's syndrome

"I first had trouble with dizziness twenty years ago. It was diagnosed by my doctor as Ménière's syndrome, but even after many tests my doctor couldn't tell me what caused Ménière's. I guess the doctors really don't know what the cause is—just like a lot of other diseases, I suppose. Anyhow, I managed to get by in spite of the frequent dizzy spells until six years ago, when I became worse. Started getting daily attacks of dizziness and vomiting. The last few years I had a roaring in my left ear along with the dizzy spells. My hearing also has gotten worse during this time. My doctors tried everything, including histamine shots, vitamins, and diuretics, to no avail. I've tried home remedies like ginger and tea; that works probably just as well as all the medicines they gave me.

"One thing the doctors found—and I'm not sure whether it is related to the Ménière's or not—was high blood pressure, elevated blood sugar, and high cholesterol. The blood sugar was not terrible, but the cholesterol was over 500 mg/dl—very high. My doctor suggested a diet. He said I should watch out for fatty meats, eggs, whole milk, and cheese. I was told to eat more vegetables as well as chicken and fish. He also prescribed Atromed-S, a drug to lower cholesterol, but he discontinued it when he found out about all the side effects. (He didn't tell me what the side effects were, but I found them out for myself by looking through the *Physician's Desk Reference*. I'll never take a drug again without checking that book.) After experimenting with various drugs, the best I could do was to decrease my cholesterol to 310 mg/dl. My doctor said nothing else could be done, and that was the end of that. Earlier this year I was hospitalized with chest pain, but fortunately there was no heart damage. However, I became allergic to the nitroglycerine patches they prescribed.

"One thing I forgot to tell you is that for the last ten years I have been impotent. That is distressing to me and my wife since we still have sexual desires in spite of our age. The doctors have not been very helpful, but I guess this is something we don't talk about in public."

Mr. Johnson was surprised that there *was* something else that could be done for him. I pointed out that eating chicken and fish prevented a further drop in cholesterol, which was a shock to him because almost every dietitian insisted that he could continue to eat these foods. But he knew from experience that his cholesterol level did not budge below 300 on this regimen, and he was very anxious to start the ten-point program.

Mr. Johnson began to follow all the recommendations exactly. He is walking five miles daily and enjoying it. The symptoms of dizziness and roaring have improved markedly. When questioned about his impotence, he smiles and agrees it is better, but he is too embarrassed to divulge many details. His blood pressure has dropped from 170/90 to 130/84. His weight has dropped from 216 to 184 and is continuing to drop. Within six months his blood sugar became normal, and his cholesterol level has dropped from 310 to 223 mg/dl.

Atherosclerosis may be responsible for decreased blood circulation throughout the body. Impotence and the symptoms of Ménières (dizziness, roaring sounds, deafness) may result from poor circulation to the genitalia and inner ear. A dietary approach to improve blood circulation should always be the treatment of choice.

Case Study: Retired sales manager
Name: Eleanor Raskin
Age: 49
Problems: Cardiac disability, high cholesterol

"I was a two-and-a-half-pack-a-day smoker from age sixteen to age forty-six, but I gave that all up three years ago after my doctor read me the riot act. Actually, I should have given it up earlier, like in 1978 when I had two heart attacks. I guess I'm a slow learner. You cannot imagine the fatigue, depression, and loss of self-esteem that go with these heart attacks. Up until recently I have had to take about ten pills a day for my heart just to barely keep up with my simple daily needs since my early retirement, but now all that has stopped.

"I was astounded by Dr. Weissman's revelation; it all made such good sense. Starting on his regimen has given me a new lease on life. I am walking five miles every day and enjoying these new foods. Can you imagine that all the doctors and specialists I saw never helped as much as I have helped myself with Dr. Weissman's program? My blood cholesterol has dropped from 224 to 176 in just a short time, and I have been able to get rid of the water pills, digitalis, and the other heart medicines. I am a believer."

Case Study: Civil service employee
Name: Roberta Kelly
Age: 38
Problems: Recurring massive hives, unsuspected
    cholesterol elevation

"Sixteen years ago I developed hives all over my body. They lasted about three months but then disappeared just as mysteriously as they came—that is, until late last year. They returned to haunt me again, but now they were more extensive and persistent than before. The itching and swelling was unbearable. Just about everything I did to try and figure out the cause was futile. I suspected I was allergic to foods like shrimp and strawberries, but cutting them out of my diet made no difference. My doctor prescribed several drugs that kept the swelling down, but I was so exhausted from them that I couldn't function well at my job.

"Finally I saw Dr. Weissman. He suggested that I eliminate foods containing artificial ingredients such as synthetic colors, flavors, sweeteners, and preservatives, among others. That seemed to be the answer, because the hives cleared and I remain free of them now, six months later. He had also discovered that my cholesterol was 248 mg/dl, and I decided to follow his entire program to take care of that problem, too. I feel great since my hives have left, and changing my diet has caused my cholesterol to drop to 174 mg/dl."

Case Study: Business manager
Name: Elsie Martin
Age: 24
Problems: Hypoglycemia, premenstrual syndrome (PMS), minimal overweight

"I gained weight slowly over a number of years. By age 24, I weighed 125 pounds, too much for my height and small frame. I had also developed symptoms of hypoglycemia (low blood sugar). If I did not eat on time, I would get shaky and irritable. This was not only uncomfortable, it was frustrating. I hated the feeling of being a slave to food. In addition, I also had symptoms of PMS—cravings for sweets, irritability, bouts of uncontrollable crying, depression, and headaches.

"I had attempted and endured almost every diet on the market—Weight Watchers, Atkins, Scarsdale, High Protein, and Beverly Hills. They all worked for a time, but whatever weight loss I managed was only temporary. Because of my hypoglycemia, my previous doctor advised me to eat chicken, tuna, and other animal proteins all day long, and this inevitably caused a weight gain.

"When the low-toxin program was first suggested to me, I

dismissed it as impractical. I was young, and such a severely restrictive diet would limit my social life. It would mean giving up almost everything I enjoyed.

"Finally, however, after continued frustration with the more popular fad diets, I agreed to try it for one week. In that short period, I saw such marked results I decided to stick to it. I lost 8 pounds, PMS became a thing of the past (except when I cheat, which is becoming less frequent), and my hypoglycemia symptoms vanished completely. I no longer had to stick to a rigid time clock for my meals. I was in control of my life and my eating habits."

Case Study: Retired naval officer, department manager
Name: Jack O'Brien
Age: 48
Problem: Diabetes and subsequent complications arising
   from treatment

Jack's case is not a low-toxin success story but a cautionary tale. Jack is a retired naval officer. He had always taken care of his health and had always had major responsibilities throughout his career. In the navy he had annual physical examinations, and no health problems ever showed up. After retiring from the navy, he was offered a position as a department manager at a high-tech company. For that job, he had to take a routine physical. The exam revealed urine and blood sugar levels that strongly indicated diabetes. He was told to check with his own doctor.

His doctor confirmed the diabetic condition and instructed him on a standard diabetic diet—not the low-toxin diet. Jack was given a prescription of Tolbutamide to lower his blood sugar and was advised to lose 15 pounds. He seemed to do well with this treatment and did manage to lose 5 pounds.

About nine months later he decided to leave his new job. He and his wife made plans to move to Australia. They had sold their house and furniture and were about to leave when suddenly Jack came down with what he thought was an infection—fever, sore throat, and a rash. Later it was discovered that this was a reaction to the Tolbutamide, but no one realized it at the time. He was treated with penicillin but developed an immediate and severe allergic reaction to it. He passed out and was swiftly admitted to a hospital intensive care unit.

He remained hospitalized for three and a half months, a period that was nightmarish to Jack, who had never known serious illness before. His rash became much worse, now involving his entire body, and his skin began to slough off. He was being treated with high doses of cortisone, and he eventually appeared to improve. But then he became paralyzed, developed hemolytic anemia, and suffered kidney failure.

After a while he was able to get out of bed, but it took him a long time to learn how to walk again. When he was finally released from the hospital, he was able to walk but had a limp. His life had completely fallen apart because of adverse reactions to drugs, prescribed for a condition that would have been better treated with diet alone. I have had many diabetic patients who have responded beautifully to the low-toxin diet in combination with exercise.

Case Study: Housewife, quality control inspector
Name: Karen Brinkman
Age: 56
Problems: Recurrent bronchial asthma and acute
   bronchitis, high blood pressure, elevated blood
   cholesterol, coronary heart disease, overweight

Karen had been overweight for a number of years. She was relatively inactive and at 5 feet 4 inches weighed 185 pounds. Her eating habits were rather typical of the average American. For breakfast, she would usually have grapefruit juice, bacon and eggs, a slice of white toast with butter and jelly, and coffee. Her lunch might be a small steak, salad with Thousand Island dressing, french fries, tomato soup, and coffee. For dinner, she would have either barbecued pork ribs or barbecued chicken, salad with Good Seasons Italian dressing, baked beans, canned fruit or ice cream, and coffee.

Before she was 45, Karen had already developed high blood pressure and elevated levels of cholesterol, triglycerides, and uric acid, so she was not very healthy. To control these conditions, she was taking three pills each of aldomet, apresoline, and hydrochlorothiazide every day, but she was not given any dietary instructions.

At age 48 she suffered her first heart attack. By 50 she had developed asthma. It was so severe she was hospitalized. Subsequently she suffered from asthma whenever she caught a cold or had a respiratory infection. She also had chest pain when under stress. Drugs were prescribed to control her asthma and chest pain; she was taking them along with her blood pressure pills.

Three months after she started the low-toxin program, eliminating meats and processed foods, drinking distilled water, and taking antioxidant vitamin supplements, her weight was down to 177 pounds and her blood pressure had improved. She has been able to cut her medication down to only three pills a day and has hopes of eventually going off those too.

Karen has difficulty adhering to the exercise program because of a lack of time for concentrated walking, but she realizes that even the little she has been able to do has helped, and she intends to remedy this. She is doing reasonably well and will do better with more exercise.

Case Study: Screenwriter
Name: Paul Jacobs
Age: 35
Problems: Former drug addict, alcoholic, and heavy
    smoker, recent surgery for testicular cancer

Paul decided to change his life six years ago. His wife had just left him, and he was fired from his writing position with a major film studio because of his drug and alcohol habits. But the most important event that precipitated the change was the death of his father from lung cancer. His father, too, had been a heavy smoker.

One year ago, after five years of staying relatively clear of his destructive habits, Paul discovered a hard mass in his left testicle that turned out to cancerous. It was removed, and an abdominal operation was done to check for metastases; there were none.

At present, Paul is following the low-toxin program, though he is taking considerably more vitamin C than I usually suggest (with cancer, there are good reasons for this). He is doing well at this time. Although it is too early to be certain that Paul is entirely out of danger, his attitude and newfound respect for his body may be able to overcome the abuse he heaped on it years ago.

Case Study: Artist
Name: Harriet M. Smith
Age: 72
Problems: Chronic hives, arthritis, overweight, high
    cholesterol

Harriet had always considered herself to be in good health. An active artist, she was able to attend art shows in her late sixties and early seventies, even though arthritis made it more difficult for her to get around than it had been when she was younger.

It was after the 1984 Thanksgiving holiday that she began to have problems. She broke out in a rash. At first she thought nothing of it, assuming that it was some sort of allergy and would go away. However, when it did not disappear after several days, she went to her doctor, who prescribed a strong antihistamine, Benadryl. That cleared the rash, but it caused her to sleep constantly, so she stopped taking it. The hives were back within a week.

She returned to the doctor, who now prescribed Medrol Dosepak, a form of cortisone. It worked; the hives and itch went away. She stopped taking it then, and within two weeks the problem returned.

After five months had passed and her condition had not improved, the doctor recommended that she come to see me. Tests revealed that her blood cholesterol was high, about 275. I was unable to determine what was causing her condition, though I suspected chemical additives in foods.

In April 1985 I recommended that she give up all processed foods and suggested that she start the low-toxin program. I put her on an exercise program of walking each day, which she resisted, especially when her arthritis was bothering her. After a week of this regimen, without any drugs, her hives cleared completely. They only return when her children come to town and take her out to eat at restaurants.

This program has also allowed her to lose weight. Within four months, she dropped from 170 to 154 pounds and continued to lose. Her blood cholesterol level fell to 187, and even her arthritis began to improve. Her energy level increased, and she was able to be more active than before. After she got used to the diet, she did not feel deprived by it.

Chapter **15**

Meal Preparation
and Recipe Ideas

The change to a low-toxin diet will require some adjustments in shopping and meal planning. Since cooking is a highly personalized endeavor, most people will want to develop their own individual approaches, but some suggestions may be helpful to ease the period of transition.

As you have gone through the ten-week program, you have been making changes in your shopping list. Use the following pages as a reminder to adjust the list, removing undesirable products and adding those allowed by the program.

COOKING TIPS

Since animal foods are to be avoided on the low-toxin diet, many people may feel that they are about to enter the Twilight Zone without their old favorites. There's no question that a major adjustment is necessary, but it's fairly easy if you take it in stages, as you've been doing for the last ten weeks. Because your selection of foods is limited, advance planning and preparation are helpful in obtaining greater variety in your meals.

Soups and sauces that require a long cooking time can be prepared in advance. Cook in bulk so that you can prepare more than one meal in the same amount of cooking time. Freeze any unused food if you do not plan to eat it within a few days. Health food stores sell egg substitutes that act as a binding ingredient for baking. "Milk" for use in baking and other recipes can be prepared by blending water with groats, brown rice, zucchini, or nonfat soy powder in a blender or food processor.

In planning your meals, it is important to include frequent weekly servings of the yellow, orange, and dark-green vegetables, which are high in beta-carotene and valuable antioxidants. Of course, these vegetables should be fresh if at all possible. You may make unlimited use of them, individually or in combination.

Build the dinner meal around a starch such as brown rice, baked potatoes, corn on the cob, or pasta. These can be complemented with any variety of vegetables, legumes, soups, sauces, etc. On the low-toxin diet you may experiment and not feel bound by tradition. If you wish to have vegetables for breakfast or steel-cut oatmeal for dinner, go right ahead. Although it is a sacrifice for most people to give up their old favorites, the health benefits that accrue are surely worth it. But make the transition gradually so that the program does not become overwhelming.

The diversity of herbs and spices available will help to add zest to plant foods. If possible, grow your own. If not, obtain them fresh. Failing that, dried commercial herbs and spices are acceptable, but buy them in small amounts and refrigerate them since they are perishable and do not retain their true flavor for very long. If a dish requires extensive cooking—such as a soup or a sauce—add the herbs and spices toward the end of the cooking time, as their flavor gets lost during prolonged heating. For leftovers, a pinch or two of appropriate herbs during warming will return flavor to the dish.

It is difficult to recommend what goes well with what since tastes vary. For example, garlic, onions, and cilantro are good in sauces and soups, and fresh ginger goes well with many dishes in addition to the traditional Oriental ones, but only if you like these seasonings. The standard spices and herbs you will find useful include the following.

allspice	anise	basil
bay leaf	caraway seed	cardamom seed
celery seed	chervil	chili powder
cilantro	chives	garlic chives
cinnamon	cloves	coriander seed
cumin seed	curry powder	dill
fennel seed	ginger	garlic
lemon grass	mace	marjoram
mint	mustard	Italian parsley
oregano	paprika	parsley
nutmeg	white pepper	black pepper
red pepper	poppy seed	rosemary
saffron	sage	savory
sesame seed	tarragon	thyme
turmeric		

In addition to these standard herbs and spices, others are available at ethnic groceries specializing in Italian, Greek, Middle Eastern, and Oriental foods.

Because of the possible link between aluminum and Alzheimer's disease, do not use aluminum cookware, utensils, or foil. Microwave ovens are perfectly safe as long as they are leakproof. All new microwaves are tested for leakage; if you have an older one, you might want to have it retested. Where possible, store leftovers in glass containers.

RECIPES

Most of the recipes that follow have no salt, few simple sugars, and far less oil than standard recipes (or no oil at all). They are recipes used by my family and were devised by myself, my wife Phyllis, and our daughter Laurie Bucher. I also have to thank my running buddy Pipo Calderone, master chef and restaurateur, owner of Viva La Pasta Restaurants. He contributed

SAMPLE MENU FOR A DAY

BREAKFAST

Grapefruit half (unsweetened)
Steel-cut oatmeal with sliced peach or banana
Whole wheat toast
Distilled water, herbal tea, or fresh orange juice
Antioxidants (vitamins)

LUNCH

Split pea soup
Green salad with lemon and herb dressing
Avocado sandwich on sourdough rolls with lettuce and to-
    mato, seasoned with Dijon mustard and rice vinegar
Distilled water, decaffeinated coffee, or tomato juice
Antioxidants (vitamins)

DINNER

Mixed vegetable salad: lettuce, tomatoes, cucumber, sprouts,
    fresh mushrooms, sweet green and red peppers
Minestrone soup
Spaghetti marinara
Crusty Italian or sourdough bread
Distilled water or pineapple juice
Antioxidants (vitamins)
Fresh fruit, in season

---

the pasta recipes and helped us adjust the seasonings and measurements for other recipes. Since Pipo is familiar with the low-
toxin program, standard vegetarian diets, and the Pritikin diet,
eating at his establishments is both enjoyable and healthy. But
even at Viva La Pasta, one must select carefully from the menu.
    Because of space limitations, I have included only a few

recipes. I am currently preparing a book of low-toxin recipes, and you can write me at the Medical Center for Health and Longevity (see Appendix B) for further information about it.

BREAKFAST

Steel-Cut Oats

3 cups steel-cut oats   *9 cups distilled water*

Bring water to a boil, add oats, and lower heat (or use a Japanese rice cooker). Cook at a low simmer for 30 to 40 minutes. Help yourself to today's breakfast portion(s) and allow the remainder to cool, then store in a glass container and refrigerate. Reheat the cereal each day as needed; 2 minutes in a microwave oven or 5 minutes in a double boiler are adequate.

Do not add sugar, salt, or milk. Fresh fruit, cinnamon, nutmeg, and/or raisins may be used on cereal if desired.

Basic Oat Bran Muffins

6 cups oat bran
1½ cups rolled oats
*2 Tbs. nonaluminum, low-
   sodium baking powder
   (Featherweight brand
   obtainable in health food
   stores fits the bill)*

¾ cup raisins
2 tsp. cinnamon (optional)
2 tsp. nutmeg (optional)
¾ cup unsulfured molasses
*4¾ cups warm distilled
   water*

Mix all dry ingredients until well blended. Add molasses and warm distilled water. Spoon mixture into nonstick muffin tins. Bake in preheated 375° oven for 40 to 45 minutes. *Makes 2 dozen muffins.*

Al Gelman's Banana–Oat Bran Muffins

2 ripe bananas
⅔ cup raisins
2 cups warm distilled water

2½ cups oat bran
2 tsp. pure vanilla

Liquify bananas, ⅓ cup raisins, and ½ cup water in a blender. Combine remaining water and raisins with oat bran, vanilla, and liquified banana mixture. Mix well. Spoon into nonstick muffin tin. Bake in preheated 375° oven for 42 minutes. *Makes 1 dozen muffins.*

SALADS

Moroccan Eggplant Salad

1 medium onion, minced
3 cloves garlic, minced
1 cup flat parsley, chopped
⅓ cup cilantro, chopped
2 eggplants, cubed
1 6-oz. can tomato paste

3 Tbs. cider vinegar
½ tsp. ground ginger
2 tsp. paprika
1 bay leaf
2 tsp. olive oil
2 cups distilled water

Sauté onion, garlic, parsley, and cilantro until onion is translucent. Add eggplant to mixture and cook until eggplant is softened slightly. Combine tomato paste, vinegar, ginger, paprika, bay leaf, olive oil, and water. Add to eggplant mixture and simmer over low heat for 40 minutes, stirring frequently. Serve hot or cold. *Serves 8–10.*

Summer Squash Salad

2 large yellow summer
  squash, sliced
juice of ½ fresh lemon
2 green onions, minced
freshly ground white
  pepper to taste

pinch of powdered hot
  mustard
¼ cup unsweetened rice or
  wine vinegar
1 Tbs. minced sweet red
  pepper

Combine ingredients and marinate overnight or for several hours during the day. Serve on a bed of lettuce. *Serves 3–4.*

Sprout and Mushroom Salad

2 cups mixed sprouts
(alfalfa, mung bean,
garbanzo, etc.)
2 Tbs. chopped chives

½ lb. white mushrooms,
finely sliced
1 tsp. chopped cilantro

Combine all ingredients and serve with lemon herb dressing.
Serves 3–4.

Pickled Beets

4 medium beets
distilled water (enough to
partially cover beets)
3 Tbs. cider vinegar
1 clove garlic, finely
minced

1 tsp. pickling spices
fruit juice sweetener to taste
(optional)
1 small Bermuda onion,
sliced into rings

Cook beets in pressure cooker with distilled water, 10 pounds
pressure for 12 minutes. Allow to cool. Remove tough outer
skin, then slice beets. Add vinegar, garlic, spices, and fruit
juice to cooking liquid; simmer for 1 minute. Cool mixture and
combine with beets. Marinate overnight in the refrigerator.
Remove beets from marinade and serve on a bed of lettuce
leaves. Garnish with onion rings. *Serves 4–5.*

Mustard Sauce

¼ cup leftover cooked
brown rice
¼ cup cold distilled water

juice of ½ fresh lemon
4 Tbs. Dijon mustard
2 tsp. chives

Combine all ingredients in blender. Excellent topping for
salads, asparagus, and other vegetables.

Three- or Four-Bean Salad

*1 cup cooked pinto or
kidney beans*
2 cups steamed green beans
*1 cup cooked garbanzo
beans*
*1 cup steamed wax beans
(optional)*
*1 cup minced bok choy or
celery*
*½ cup minced sweet red or
green pepper*

1 Bermuda onion, sliced
*1 tsp. cold pressed
safflower oil*
½ tsp. sesame seed oil
juice of 1 fresh lemon
½ cup cider vinegar
¼ cup honey
3 cloves garlic
*⅛ tsp. freshly ground
pepper*

Place all vegetables in a large bowl. Blend the oil, lemon juice, vinegar, honey, garlic, and pepper in a blender until garlic is liquefied. Add to vegetables, mix well, and refrigerate for at least 2 hours. *Serves 8–10.*

SOUPS

Phyllis's Favorite Split Pea Soup

*1 lb. yellow or green split
peas*
2 quarts distilled water
2 bay leaves
1 carrot
2 cloves garlic, minced

*1 large yellow onion,
quartered*
2 celery stalks
¼ tsp. pepper
⅛ tsp. thyme
¼ tsp. oregano

Add peas to boiling distilled water, then add remaining ingredients, saving thyme and oregano until the last 15 minutes. Simmer slowly for 2 hours. Allow to cool. Remove bay leaves. Purée cooled soup in a blender or food mill. Reheat and serve. *Serves 6–8.*

Croutons

½ loaf sourdough, crusty French, or Italian bread

Remove bread crust, discard, and cut bread into half-inch cubes. Toast cubes on a cookie sheet in a 350° oven until golden (approximately 15 minutes). Save in covered container or serve with soup or salad. *Makes 2 cups.*

L-T Vegetable Soup

2½ quarts distilled water or stock saved from steamed vegetables
2 medium boiling potatoes, quartered
2 carrots, sliced
1 large yellow onion, sliced
1 cup cauliflower florets
1 cup broccoli florets
1 parsnip, sliced

4 ripe medium tomatoes, quartered
1 cup chopped bok choy
1 Tbs. finely chopped fresh cilantro (or ⅛ tsp. powdered coriander)
2 tsp. chopped fresh oregano (or ¼ tsp. dried oregano flakes)

Add potatoes and carrots to boiling distilled water or stock. After 5 minutes add remaining vegetables and simmer over low heat until vegetables are done but still somewhat firm (about 15 minutes). *Serves 6–8.*

Minestrone

2 quarts distilled water
2 cups cooked pinto, adzuki, kidney, or black beans (or a mixture)
1 cup diced carrots
1 medium zucchini, diced
1 cup broccoli
1 cup shredded green or Savoy cabbage
1 leek, shredded (or 1 yellow onion, diced)
2 pounds chopped plum tomatoes (or 1 small can salt-free tomato paste)

3 Tbs. finely chopped cilantro
1½ Tbs. finely chopped fresh basil
¼ tsp. freshly ground white pepper
5 cloves garlic, minced
6 sprigs flat (Italian) parsley, finely minced
⅛ tsp. thyme
¾ cup whole wheat elbow macaroni

Combine water, precooked beans, chopped vegetables, and seasonings in a large kettle; simmer 30 minutes. Add macaroni and simmer an additional 15 minutes, or until pasta is tender. Add more water during the cooking process if necessary. Serve with crusty Italian bread. *Serves 8–10.*

Gazpacho

1½ quarts tomato juice (avoid reconstituted and salted juice if possible)
juice of 3 fresh lemons or limes (or ½ cup apple cider vinegar and juice of ½ fresh lemon)
1 Tbs. finely chopped fresh oregano
2 cloves garlic, minced
⅓ New Mexico medium-hot chili pod, finely chopped or ground
black pepper to taste
hot red chili pepper to taste (optional)

chopped chives (optional)
2–3 large beefsteak tomatoes, peeled and chopped (or 1 28-oz. can of ready-cut peeled tomatoes)
1 large green pepper, seeded, cored, and finely chopped
2 medium zucchini or cucumbers, coarsely chopped
1 small Bermuda onion, finely chopped (optional)

Use a blender to combine juices, oregano, garlic, seasonings, and part of the vegetables. Do not liquefy all the vegetables; at least half should be added in the coarsely chopped state. Refrigerate for several hours and serve chilled in cups or soup bowls. The optional seasonings add zest for those who like their gazpacho spicier; you can also serve them separately and let everyone add them as desired. *Serves 8–10.*

VEGETABLE ENTRÉES

Steamed vegetables are a staple of the low-toxin diet. To add flavor to them, you may want to use the following mixture in your vegetable steamer.

1 cup distilled water	1 Tbs. chopped cilantro
1 Tbs. chopped parsley	1 Tbs. chives

Vegetables have different cooking times. If you plan to have several vegetables at the same meal, start the ones that take longer to cook first and then add the ones requiring less time. We like our vegetables crunchy, so if you follow the recommendations in Table 15.1, the vegetables may seem a bit undercooked; just cook them a little longer if you like, but don't overdo the steaming. Don't forget to save the steaming liquid for stock. Steaming racks come with many pressure cookers or can be purchased for use with any large stainless steel pot.

TABLE 15.1
Vegetable Steaming Times

10–15 minutes	5–7 minutes	5 minutes or less
artichokes	asparagus	bean sprouts
Brussels sprouts	bell peppers (red,	bok choy (Chinese
carrots	yellow and	celery)
celery	green)	Chinese or Napa
eggplant	broccoli (cut)	cabbage
onions	cabbage (cut)	collard greens
whole okra	cauliflower (cut)	endive leaves
stringbeans	corn on the cob	kale
fava beans	leeks	mushrooms
	snow peas	mustard greens
	summer squash	spinach
	yams (½″ slices)	sugar snap peas
	yellow squash	Swiss chard
	zucchini	tomatoes

Steamed vegetables are simplicity itself and should be served with rice or another whole grain or a baked potato.

Some root vegetables and winter squash need to be steamed longer than 15 minutes or will actually do better with boiling. If you steam them, slice them first. And, of course, you can always eat your vegetables raw, even those that are traditionally cooked; for example, corn on the cob, if it is really fresh, tastes great uncooked.

Steamed Curried Vegetables

4–5 vegetables, your choice
  (carrots, spinach, leeks,
  zucchini, eggplant, etc.)
1 tsp. curry powder
1 Tbs. cornstarch
3 cups distilled water

4 green onions, chopped
2 cloves garlic, minced
¼ tsp. paprika
⅛ tsp. turmeric
1 Tbs. chopped cilantro
⅛ tsp. mustard powder

Steam the vegetables.

Dissolve curry powder and cornstarch in ½ cup water. Combine remaining water, vegetables, and all other ingredients and simmer gently over low heat for 5 minutes. Add curry and cornstarch to mixture and stir until thickened. May be used as a sauce for grains or beans. *Makes 3 cups.*

Hot-and-Sour Stir Fry

5 dried Chinese
  mushrooms, soaked and
  softened in 1 cup
  distilled water
juice of 1 lemon, strained
2 Tbs. unsweetened rice
  vinegar
3 Tbs. dry Sauterne
3 Tbs. low-sodium soy or
  tamari sauce
  (unpreserved)
⅛ tsp. turmeric

1 Tbs. arrowroot or
  unbleached cornstarch
¼ tsp. ground chili pepper
  (or to taste)
1 Tbs. safflower oil
3 cloves garlic, minced
¾" cube fresh ginger root,
  peeled and minced
1 onion, finely chopped
3 medium zucchini, sliced
2 cups bok choy, chopped

Remove mushrooms from water and slice into thin slivers, reserving water; discard tough stems.

Mix lemon juice, rice vinegar, wine and soy sauce. Add turmeric, arrowroot, and chili pepper to this mixture and stir until arrowroot is dissolved.

Place a wok over high heat. Add oil and heat 30 seconds. Fry garlic and ginger quickly, being careful not to burn. Add onion and fry quickly, then add zucchini and bok choy, stirring until vegetables are hot. Push vegetables to the side, add mushroom liquid directly to wok, and heat to boiling. Add lemon juice mixture and blend with vegetables. Cover and cook 7 minutes, stirring occasionally. Serve with steamed brown rice or pasta. *Serves 6–8.*

Vegetable Lo Mein

½ lb. whole wheat spaghetti or buckwheat soba noodles

5 dried Chinese mushrooms, soaked for at least 1 hour in 1 cup distilled water

2 Tbs. dry Madeira wine

3 Tbs. low-sodium soy or tamari sauce (unpreserved)

2 Tbs. cornstarch

⅛ tsp. ground white pepper

2 tsp. sesame oil

2 Tbs. peanut or safflower oil

1 onion, minced

4 cloves garlic, finely minced

1 thick slice eggplant, cut into bite-sized pieces

1 medium carrot, steamed and cut into bite-sized pieces

¼ pound snow peas, ends removed

1 cup thinly sliced bamboo shoots

2 cups bok choy, finely sliced

1 cup mung bean sprouts

4 green onions, chopped, for garnish

Cook spaghetti *al dente*, (7 to 10 minutes), drain, and set aside.

Remove mushrooms from water and slice into thin slivers, reserving water; discard tough stems. Add wine, soy sauce, cornstarch, pepper, and sesame oil to reserved water; stir until cornstarch is dissolved.

Place a wok over high heat. Add peanut oil and heat 30 seconds. Fry onion and garlic quickly (5 to 10 seconds), being careful not to burn. Add eggplant and stir until softened. Add remaining vegetables and stir until vegetables are hot. Push vegetables to the side, add mushroom liquid directly to wok, heat to boiling, and stir in vegetables. Cover and cook about 5 minutes, stirring occasionally. Add noodles, mix until heated through, and garnish with green onions. *Serves 4–6.*

Ratatouille

1 eggplant, cubed
3 medium zucchini, sliced
2 onions, chopped
2 red bell peppers, roasted, seeded and sliced
2 large tomatoes, skinned and chopped.

3 cloves garlic, minced
1 Tbs. minced fresh basil
1 Tbs. minced fresh coriander
⅛ tsp. ground pepper
1 Tbs. chopped chives, for garnish

Cook all ingredients in a covered pot, simmering gently until vegetables are tender. Garnish with chives and serve hot or cold with brown rice or pasta. Tastes better after reheating. *Serves 6–8.*

ENTRÉES

Onion Sauté and Kashi

2 cups Kashi
4⅓ cups distilled water
½ onion, finely chopped
1 clove garlic, minced
½ tsp. minced cilantro

1 tsp. minced fresh oregano
1 Tbs. safflower oil
3 Tbs. rice wine
3 Tbs. tamari

Add Kashi (a blend of oats, brown rice, rye, wheat, triticale, buckwheat, barley, and dehulled sesame seeds) to 4 cups boiling water. (If Kashi is not available, you can make a reasonable substitute by mixing whatever assortment of grains you have at home.) Cover tightly and steam over medium heat 20 to 25 minutes or until all water is absorbed. Remove from heat.

Sauté onion, garlic, cilantro, and oregano in oil until tender. Add remaining water, wine, and tamari; simmer for 5 minutes, then add to Kashi mixture. Heat in oven if necessary. Excellent with steamed vegetables. *Serves 8.*

Wild Rice and Bulgur Wheat Pilaf

½ cup wild rice
½ cup bulgur wheat
3 cups vegetable broth or
  distilled water
4 green onions, minced
¼ cup raisins

2 tomatoes, chopped
½ tsp. sage
1 tsp. chopped parsley
¼ tsp. allspice
½ cup chopped pine nuts
  or cashews

Combine all ingredients except nuts in a Japanese rice cooker and heat until liquid is absorbed (30 to 45 minutes). Add nuts and heat for 2 additional minutes. *Serves 4–6.*

PASTA ENTRÉES

Capellini alla Checca

1 lb. durum wheat angel
  hair pasta (or your
  preferred type of pasta)
3 cloves garlic, finely
  minced

1 Tbs. olive oil
6 sweet basil leaves,
  chopped
1 pound ripe tomatoes,
  skinned and chopped

Boil pasta *al dente* in distilled water (angel hair cooks quickly, in 3 to 4 minutes; regular spaghetti cooks in about 9 minutes).

Sauté garlic in oil until translucent. Add chopped basil and cook briefly, about 2 minutes. Add garlic and basil to the raw tomatoes and combine. Mix well with hot pasta and serve immediately (or heat briefly in a microwave oven if desired). *Serves 4.*

A nice variation on this recipe is to add fresh young asparagus to the sauce. Use ½–1 lb. and cut the upper part of the stalks into ½″ pieces. Stir-fry them in 1 Tbs. olive oil until tender-crisp and bright green before adding.

Marinara Sauce

*3 lbs. Italian plum
tomatoes, skinned and
chopped
1 small onion, minced
1 Tbs. olive oil
4 cloves garlic, minced
2 Tbs. chopped fresh
oregano (or 1 tsp. dry
oregano flakes)*

*10 sweet basil leaves,
chopped
1 tsp. apple juice
½ tsp. marjoram
⅛ tsp. nutmeg
⅛ tsp. cayenne pepper
(optional)
3 Tbs. dry Sauterne or
Chianti (optional)*

Simmer chopped tomatoes slowly for 2 or more hours until sauce thickens. Separately sauté onion in oil until translucent and tender, then add garlic and sauté an additional 2 minutes. Add to tomato mixture after it has thickened. Add remaining ingredients and cook another 15 minutes. Serve over spaghetti cooked *al dente*. *Makes 5½ cups*.

Marinara Sauce with Mushrooms

Sauté ¼–½ lb. sliced fresh mushrooms with the garlic and onions in the basic recipe. Dried Italian porcini mushrooms may be used as a special treat, but do not sauté them. Reconstitute them with water and add to the sauce just prior to serving.

Part III

The Research

Chapter 16

The Myths of
the Modern Diseases

Many of us can look back to ancestors who lived productively into old age in the early part of this century, untouched by cancer, coronary heart disease, or diabetes. Yet many of today's older generation barely survive in countless nursing homes, warehoused with debilitating "degenerative" diseases. Many of them are dying younger than their parents or grandparents.

They are suffering and dying from diseases unheard of before the modern era, or from conditions that were known but extremely rare. They—and we—are led to believe that many of their ailments are merely a part of the aging process, unavoidable as bodies "degenerate," but in almost all instances the causes of these "modern" maladies have remained obscure. However, virtually all of them are relatively new in human history. With the exception of a few, like cancer and diabetes, most have developed within the last 200 years.

Scientists have spent billions on research to find the causes of each of these major diseases, but with limited success. In almost all cases, a specific cause has eluded them. A virus is often suspected but rarely found. I believe that a new approach is needed. The modern diseases are *not* "degenerative." They have *not* been with us always. And they may well be interrelated.

New research needs a revised focus. Medical science and the public can no longer ignore the historic development of the major modern killers—for they most definitely are diseases that have developed and spread dramatically in recent history (see Table 16.1).

I have already touched on these points in Part I. What I present here is a more detailed history of a few of today's leading health problems, with additional evidence to the effect that we are dealing with new phenomena linked to emerging industrialization. Even the diseases that arose earlier in history may very well have been associated with xenobiotics. For example, gout among the Romans and the early English aristocracy has been associated with lead poisoning resulting primarily from their consumption of fortified wines. Myasthenia gravis, discovered in 1672 by Sir Thomas Willis, may ultimately be found to be caused by a toxin.

INFECTIOUS DISEASES

The decline of various major infections can be correlated with the emergence and spread of the Industrial Revolution (see Tables 16.2–6). As indicated in Chapter 1, the decrease in deaths from infection did not result from medical breakthroughs, better nutrition, or improved sanitation. I maintain that the first victims of the Industrial Revolution were the viruses and bacteria that caused human disease.

I also believe that in recent years the debilitating effect of xenobiotics on the human immune system has paved the way for the emergence of various new infectious diseases such as herpes, Epstein-Barr syndrome, hepatitis, viral leukemias, and acquired immune deficiency syndrome (AIDS). These new diseases have adapted to our polluted environment far more successfully than their human victims. Although the ten-point program will not necessarily "cure" them, it is clear that the immune system needs help. Until the necessary vaccines are

TABLE 16.1
Discovery Dates of Some Noninfectious Modern Diseases

Disease	Date	Discoverer
Diabetes mellitus	1673	Thomas Willis
Parkinson's disease	1817	James Parkinson
Scleroderma	1817	Jean Louis Alibert
Hay fever	1819	John Bostock
Lupus of the skin	1828	Laurent Théodore Biett
Muscular dystrophy	1830	Charles Bell
Hodgkin's disease	1832	Thomas Hodgkin
Multiple sclerosis	1838	R. Carswell, J. Cruveilhier
Leukemia	1845	John Hughes Bennett
Neuromuscular atrophy	1849	Guillaume Duchenne
Rheumatoid arthritis	1859	Alfred Baring Garrod
Cerebral palsy	1862	William John Little
Dermato/polymyositis	1863	P. Hepp
Down's syndrome	1867	John Langdon Down
Amyotrophic lateral sclerosis	1868	Jean Martin Charcot
Systemic lupus erythematosus	1873	Moritz Kaposi
Kaposi's sarcoma	1874	Moritz Kaposi
Ulcerative colitis	1875	S. Wilks, W. Moxon
De la Tourette syndrome	1885	Gilles de la Tourette
Charcot-Marie-Tooth disease	1886	J. Charcot–P. Marie–H. Tooth
Spinal muscular atrophy	1890	G. Werdnig E. Hoffmann
Myasthenia gravis	1890	W. Erb (T. Willis, 1672)
Reading disorder of childhood	1896	N. Pringle Morgan
Sarcoidosis	1899	Caesar Boeck
Alzheimer's disease	1907	Alois Alzheimer
Myotonic dystrophy	1909	Hans Steinert
Coronary artery disease	1912	James Herrick
Reiter's syndrome	1916	Hans Reiter
Endometriosis	1927	John A. Sampson (K. Rokitansky, 1865)
Premenstrual syndrome (PMS)	1931	Robert Frank
Crohn's disease (ileitis)	1932	B. Crohn–L. Ginzburg–G. Oppenheimer
Cystic fibrosis	1936	Guido Fanconi
Kawasaki's disease	1967	Tomisaku Kawasaki

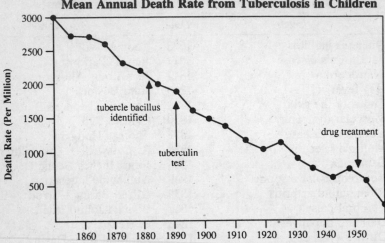

TABLE 16.2
Mean Annual Death Rate from Tuberculosis in Children

Note: Tables 16.2–6 reprinted by permission of the University of Chicago press from E. H. Kass, "Infectious Diseases and Social Change," *Journal of Infectious Diseases* 123 (1971): 110–114.

TABLE 16.3
Mean Annual Death Rate from Measles in Children

TABLE 16.4
Mean Annual Death Rate from Whooping Cough in Children

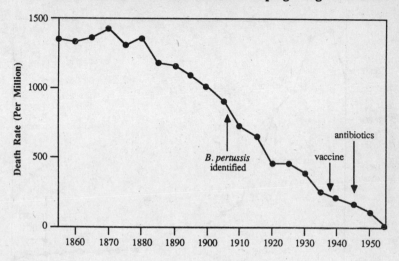

TABLE 16.5
Mean Annual Death Rate from Diphtheria in Children

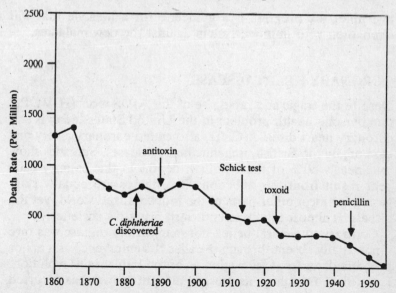

TABLE 16.6
Mean Annual Death Rate from Scarlet Fever in Children

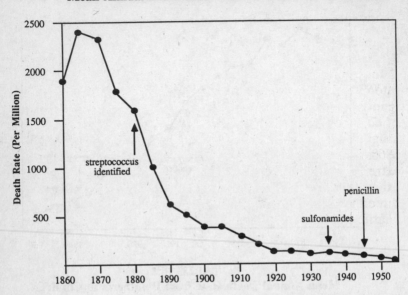

available, the program is a guideline for a lifestyle that will strengthen your immune system against the new maladies.

CORONARY HEART DISEASE

Despite the tragic and rapid rise of the AIDS virus (HIV), the number one health problem in the United States today is still coronary heart disease (CHD, also called coronary artery disease or, by the British, ischemic heart disease). Statistics show that nearly 50% of the 2 million deaths in this country each year result from this affliction. The figures are equally staggering for most other parts of the industrialized world, yet the disease is almost totally unknown in primitive societies.

Contrary to general belief, coronary artery disease was rare if not totally absent through the ages. Its emergence as a major problem dates from the beginning of the Industrial Revolution. Even more troubling, its most significant growth has occurred within this century. Though the statistics are clear, physicians and the general public alike continue to believe that it has been a major disease since the beginning of time. The view of Dr. Howard B. Sprague is typical; he stated in 1966, "Certainly the disease did not suddenly leap into existence about 1920, fully armed for destruction like Athena from the brow of Zeus."

Indeed, it did not "suddenly leap." It insidiously crawled into existence around 1910. It was so slow in evolving that we ignored the fact that a new epidemic was starting. Heart disease of *all* types—even excluding coronary heart disease—was not a significant factor in the mortality rate even as recently as the 1860s (see Table 16.7). By 1920 coronary heart disease was still extremely rare, but thereafter it began a dramatic rise. For example, while the population of Britain over age 50 increased threefold from 1901 to 1962, the number of deaths from coronaries increased 200 times.

TABLE 16.7
The Ten Leading Causes of Death

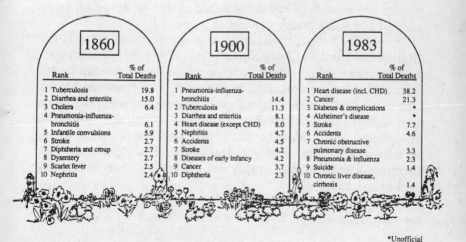

1860 Rank	% of Total Deaths
1 Tuberculosis	19.8
2 Diarrhea and enteritis	15.0
3 Cholera	6.4
4 Pneumonia-influenza-bronchitis	6.1
5 Infantile convulsions	5.9
6 Stroke	2.7
7 Diphtheria and croup	2.7
8 Dysentery	2.7
9 Scarlet fever	2.5
10 Nephritis	2.4

1900 Rank	% of Total Deaths
1 Pneumonia-influenza-bronchitis	14.4
2 Tuberculosis	11.3
3 Diarrhea and enteritis	8.1
4 Heart disease (except CHD)	8.0
5 Nephritis	4.7
6 Accidents	4.5
7 Stroke	4.2
8 Diseases of early infancy	4.2
9 Cancer	3.7
10 Diphtheria	2.3

1983 Rank	% of Total Deaths
1 Heart disease (incl. CHD)	38.2
2 Cancer	21.3
3 Diabetes & complications	*
4 Alzheimer's disease	*
5 Stroke	7.7
6 Accidents	4.6
7 Chronic obstructive pulmonary disease	3.3
8 Pneumonia & influenza	2.3
9 Suicide	1.4
10 Chronic liver disease, cirrhosis	1.4

*Unofficial

But don't historical records show that hundreds of years ago people suffered chest pains, presumably "heart attacks," and died? Yes, there are a few reports of "angina," but only a few, and other evidence suggests that the symptoms were most likely due to something besides coronary artery disease.

One piece of information often cited to "prove" that coronary heart disease has been around for thousands of years is a report written in the early part of this century by researchers who examined non-museum-grade Egyptian mummies from around 2500 B.C. However, a careful study of the report reveals that the findings are not conclusive. Since the Egyptian embalmers removed all internal organs from the bodies, it was impossible for the researchers to examine the hearts. The paper written by Sir Marc Armand Ruffer in 1911 was based upon an examination of the arms and legs and part of the aorta, the main artery from the heart. He found evidence of calcified large arteries, and that was all. Another examination by Dr. Allen R. Long in 1931, indicated that one mummy out of the many available for examination showed some evidence of thickened, calcified coronary arteries (but no evidence of narrowing, plaques, or thrombosis). Moreover, this mummy showed many signs of infection: heart valve inflammation, kidney disease, and tuberculosis—hardly a healthy person to begin with. None of the reports presented clear-cut evidence of the widespread presence of coronary artery disease in ancient Egypt.

One of the best early sources of medical information is the anatomical work of Leonardo da Vinci (1452–1519), who was, among many other things, a pioneer of heart anatomy. He dissected a large number of bodies and paid considerable attention to detail in his anatomical drawings, also keeping notes on his findings. He never mentions anything that could be interpreted as a diseased atheromatous coronary vessel. Since he was such a keen observer, this silence is a significant one: coronary artery disease simply did not exist or was extremely rare at the time.

Many in medicine regard Sir William Heberden's lecture to the Royal College of Physicians in London in 1768 as the first classic description of angina pectoris (chest pain). However, the evidence that Heberden was describing chest pain associated with coronary heart disease is far from conclusive. In the absence of autopsies, Heberden carefully avoided theorizing about the origin of the angina. "What the particular mischief is," he cautioned, "is not easy to guess, and I have had no opportunity of knowing with certainty."

In 1772 Heberden had his first and only opportunity to have an autopsy performed upon a victim of recurring chest pain and sudden death. After a careful examination of the chest and its contents, "no manifest cause of his death could be discovered," and Heberden was unable to make a confirmatory diagnosis. Although it is very possible that the angina described by Heberden was caused by coronary heart disease, it may also have been caused by syphilitic heart disease or an ailment arising from some other infection.

There are a great many infections that can cause symptoms similar or identical to those of coronary heart disease, notably bubonic plague, diphtheria, typhoid fever, pneumonia, tuberculosis, rheumatic heart disease, rheumatic fever, syphilis, scarlet fever, influenza, typhus, and pericarditis. Due to antibiotics, among other things, these infectious ailments have subsided in this century and are infrequent causes of chest pain today.

Throughout most of the nineteenth century, with all the medical advances that took place, little new information came to light about angina or coronary artery disease. Even Rudolph Virchow, who wrote the classic medical textbook of the mid-nineteenth century and was one of the foremost medical authorites of his day, did not mention the disease. Not until the end of the nineteenth century, almost 100 years after Heberden, did such physicians as George Dock, Adam Hammer, and Sir William Osler begin to report a rising incidence of angina cases. Also, during that same period, European autopsy pathologists observed an increasing number of victims with diseased coronary vessels.

The first report of prolonged chest pain, blockage of a coronary artery, and infarction, or death of heart muscle tissue, resulting in a "heart attack" was published in 1910 by Drs. W. P. Obrsastzow and N. D. Straschesko of Kiev, Russia. Two years later Dr. James B. Herrick described the disease, but it was still so rare that his article was ignored by physicians the world over. In his autobiography, *Memories of Eighty Years* (1949), he recalled humorously, "The publication aroused no interest. It fell like a dud. Recognizing the radical nature of the view I held, . . . I doggedly kept at the subject. In 1918 I showed slides and electrocardiograms. . . . physicians in America and Europe finally woke up to the diagnosis . . . translated by the layman into 'heart attack.' "

At about the same time, Dr. Paul Dudley White—later a renowned cardiologist and personal physician to President Dwight D. Eisenhower—was a young intern at Massachusetts General Hospital in Boston. He also reported years afterward that the disease was rare when he began his career. Going over his records for 700 male patients, a great many of them over 60 years of age, he found only 4 during the period 1912–13 who had coronary artery disease. It was not until 1921, when Dr. White was in private practice for two years, that he saw another patient with this condition. Year by year thereafter, the number of cases rose rapidly in the industrialized world.

Deaths from coronary artery disease were first reported in U.S. statistics in 1933. That year, there were 37.8 deaths from coronary heart disease per 100,000 people. By 1963 the figure had skyrocketed to 244 per 100,000. Since then the annual death rate from heart attacks has declined, but it still maintains its dubious first-place honor, with about half of all deaths attributable to it.

Many heart attacks occur "silently," so that the victim is not aware of a problem and does not seek medical aid. Ultimately that person may experience chest pain (angina) or another heart attack (maybe with warning, maybe not) or may die unexpectedly. Sudden cardiac death can also occur without any previous indication of heart disease. Nearly 25% of the 1.5 million annual heart attacks in the United States do not offer the victim a second chance; each year, more than 350,000 Americans fall into this category.

Autopsy studies done on supposedly healthy 18- to 25-year-old Korean-era and Vietnam battle casualties found that many of these young men already showed a 50% narrowing of their coronary arteries. Other studies have confirmed that by age 30 many American males are at high risk of a heart attack.

Physicians have determined that a very high percentage of people who experience heart attacks have certain habits and health characteristics in common (see Table 16.8). From the statistical evidence, they have isolated various "risk factors" and classified them as "major" and "minor." By using these risk factors, they can determine a patient's potential susceptibility to coronary heart disease. By eliminating those factors

TABLE 16.8
Three Major Coronary Risk Factors

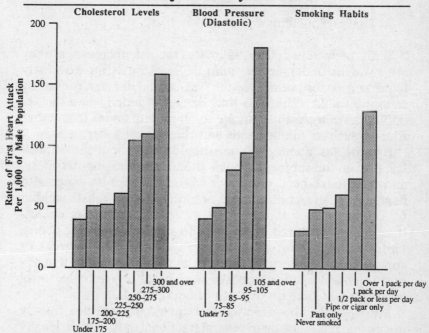

Source: National Heart, Lung, and Blood Institute, National Institutes of Health.

that are in their control, people can decrease their chances of developing the disease.

Major Risk Factors
1. Smoking
2. Elevated blood cholesterol levels
3. High blood pressure

Minor Risk Factors
1. Male sex
2. Obesity
3. A family history of heart attacks
4. Stress, combined with a Type A personality (the aggressive, driving individual)
5. Gout with arthritis
6. Postmenopause
7. Diabetes mellitus

With the factors defined as major, the risk decreases as they are brought under better control. Smoking is an extremely important factor because of the free-radical damage caused by cigarette smoke. The risks may be compounded when the xenobiotics from cigarette smoke combine with toxins from other sources such as animal foods and chlorinated water.

By now, just about everyone should know about cholesterol. It has been the subject of much attention in the popular press, and there have been numerous warnings about its dangers to health. Actually, cholesterol is not some foreign substance that we obtain entirely through diet. A certain amount of cholesterol is manufactured in the liver to supply the body's needs. Unfortunately, most of us ingest a considerable amount of contaminated cholesterol through our diets, creating dangerously high levels in the bloodstream. These elevated cholesterol levels greatly increase our risk of heart attack.

Most people who are aware of cholesterol are not equally aware that it is made up of several fractional parts—high density lipoproteins (HDL), low density lipoproteins (LDL), and very low density lipoproteins (VLDL). Research suggests that HDL cholesterol is a protective factor important in preventing heart attacks, while the others, LDL and VLDL cholesterol, are the villains. It is possible that vitamins C and E, niacin, and exercise, may raise HDL levels and lower total cholesterol levels, but a diet low in animal tissue, cholesterol, and saturated fat is also needed to keep HDL and total cholesterol at healthy levels.

High blood cholesterol levels are like red flags warning us that the body is storing too many toxins. Dietary fat and cholesterol act as carriers of xenobiotics. We cannot blame fat and cholesterol alone, for there are far too many examples of meat-eating societies that remain free of coronary artery disease and other health problems in spite of a high-fat diet.

Blood pressure is the most easily obtainable clue to risk of coronary heart disease. It is possible to prevent high blood pressure (hypertension) if efforts are made early enough, and such efforts can in turn reduce the risk of stroke and heart attack. Since 1970 various programs for lowering blood pressure have reduced the death rate from coronary heart disease and stroke in the United States by 33% and 44% respectively.

A very effective way to reduce high blood pressure is to lower total blood cholesterol levels. A more controversial approach to reducing hypertension is to eliminate salt in the diet. Sodium has long been considered the main cause of high blood pressure, but recent studies cast some doubt on this view and suggest that other substances, such as chlorides, may also be important. Not everyone is adversely affected by salt. However, until all the evidence is in, those who are salt-sensitive and have high blood pressure or a family history of high blood pressure should stay on a low-sodium or low-salt diet. And it is not enough to get rid of the salt shaker. Most Americans consume in excess of 3 grams of salt each day from processed and restaurant foods.

Among the other substances that may raise blood pressure are cadmium and chlorine (found in tap water), alcohol (high blood pressure—160/90 and above—is four times more prevalent in drinkers who consume four or more alcoholic beverages a day), and elevated lead levels in blood (partially for this reason, the Environmental Protection Agency has been attempting to terminate the use of lead in gasoline). Calcium and magnesium may lower blood pressure.

It has been determined that the danger of heart attack tends to increase with the number of risk factors an individual has. With many risk factors, the increase is a dramatic one. For example, if a hard-driving man comes from a family whose male members have all died of heart attacks at an early age, he has a strong chance of developing coronary heart disease as well. However, he should not accept that as his fate. By undertaking a concerted preventive effort early enough, he can improve his health and extend his lifetime beyond that of his ancestors.

The ten-point low-toxin program can be most effective in reducing the risk of coronary heart disease and sudden cardiac death.

CANCER

The noninfectious killer disease with the longest known history is cancer. Bone tumors were found in a few non-museum-grade Egyptian mummies, and tumors were described briefly in *Eber's Papyrus*. However, cancers of the lung, breast, stomach, colon, and lymph system, so common today, were strikingly absent in the mummies examined. Malignant cancers, though rare, also existed in ancient Greece and were described in several short paragraphs by Hippocrates in his *Aphorisms* (c. 400 B.C.). The Romans reportedly suffered from bone cancer, which recent historians have blamed on lead in their wine and drinking water, contaminated while being stored or transported in lead containers or pipes.

However, cancer was not widespread in any of the ancient civilizations.[1] In fact, there does not appear to have been a

[1]One notable exception was found in China, where for hundreds of years people living in the delta area of the southern province of Canton had a high incidence of nasopharyngeal cancer. This phenomenon has been attributed to a traditional staple of the local diet, dried salted mackerel. The dead mackerel were often left in barrels of salt water for a day or more before preservation, and the salting and drying of decaying fish resulted in high levels of nitrosamines, known carcinogens.

significant incidence of cancer in the population of any Western country until the nineteenth century.

The first cases of environmental or occupational cancer were reported by Sir Percival Pott in 1775. He was not aware that polycyclic hydrocarbons and arsenic are very common ingredients in chimney soot, but he did link the soot itself to the scrotal cancers he had observed in English chimney sweeps.

In the nineteenth century, scientists began to realize that environmental exposure to chemicals—through inhalation, skin contact, or ingestion—could result in cancers. Workers exposed to arsenic fumes in the copper and tin foundries of Cornwall and Wales developed skin malignancies. Miners of cobalt and uranium in Saxony and Bohemia suffered from a respiratory condition that turned out to be lung cancer. And industrial workers exposed to lignite, coal tar, and other chemicals began to develop cancers.

Research into cancer treatment began in the nineteenth century. Scientists Johannes Müller and Rudolph Virchow recognized the abnormality of cancer cells and suspected that it resulted from chemical irritation. In 1895, at the Johns Hopkins School of Medicine, Dr. William S. Halstead first used radical surgery to treat cancer. Further developments came that same year with the discovery of X-rays by Wilhelm Roentgen, and three years later with the discovery of radium by Marie and Pierre Curie.

Not surprisingly, during the early years of the Industrial Revolution cancer was a rare disease. In 1800 the cancer death rate was 2.1 per 100,000 people, or less than 1% of all deaths. One hundred years later the rate had climbed to 82.8 per 100,000 (about 3.7% of all deaths), and today it is much higher, approximately 180 per 100,000 (22% of all deaths).

Cancer statistics for the nineteenth century would be all but forgotten were it not for my discovery of rare books written in the early part of this century. In 1908 Dr. W. Roger Williams published *The Natural History of Cancer*, tracing the emergence of the disease. Table 16.9, which shows the prevalence of cancer in England and Wales, is from his study. Table 16.10, which comes from Dr. F. L. Hoffman's *The Mortality of Cancer*

TABLE 16.9
The Prevalence of Cancer and Its Increase
in England and Wales

Year	Total Population	Total Deaths	Cancer Deaths	Death Rate Per Million	Proportion to Population	Proportion to Total Deaths
1840	15,730,813	359,687	2,786	177	1 : 5646	1 : 129
1850	17,773,324	368,995	4,966	279	1 : 3579	1 : 74
1855	18,829,000	426,646	6,016	319	1 : 3129	1 : 70
1860	19,902,713	422,721	6,827	343	1 : 2915	1 : 62
1865	21,145,151	490,909	7,922	372	1 : 2670	1 : 62
1870	22,501,316	515,329	9,530	424	1 : 2361	1 : 54
1875	24,045,385	546,453	11,336	471	1 : 2121	1 : 48
1880	25,714,288	528,624	13,210	502	1 : 1946	1 : 40
1881	25,974,439	491,937	13,542	520	1 : 1918	1 : 36
1882	26,413,861	516,654	14,057	532	1 : 1879	1 : 36
1883	26,770,744	522,997	14,614	546	1 : 1763	1 : 35
1884	26,922,192	530,828	15,192	564	1 : 1772	1 : 35
1885	27,220,706	522,750	15,560	572	1 : 1749	1 : 33
1886	27,522,532	537,276	16,243	590	1 : 1694	1 : 33
1887	27,827,706	530,758	17,113	615	1 : 1626	1 : 31
1888	28,136,258	510,971	17,506	621	1 : 1607	1 : 29
1889	28,448,239	518,353	18,654	656	1 : 1525	1 : 30
1890	28,762,287	562,248	19,433	676	1 : 1480	1 : 28
1891	29,081,047	587,925	20,117	692	1 : 1445	1 : 29
1892	29,405,054	559,684	20,353	690	1 : 1443	1 : 27
1893	29,731,100	569,958	21,135	711	1 : 1407	1 : 27
1894	30,060,763	498,827	21,422	713	1 : 1403	1 : 23
1895	30,383,047	568,997	22,945	755	1 : 1324	1 : 24
1896	30,717,355	526,727	23,521	764	1 : 1306	1 : 22
1897	31,055,355	541,487	24,443	787	1 : 1270	1 : 22
1898	31,397,078	552,141	25,196	802	1 : 1246	1 : 22
1899	31,907,762	581,799	26,325	825	1 : 1212	1 : 22
1900	32,261,013	587,830	26,731	828	1 : 1207	1 : 22
1901	32,621,263	551,585	27,487	842	1 : 1186	1 : 20
1902	32,997,626	535,538	27,872	844	1 : 1183	1 : 19
1903	33,378,338	514,028	29,089	872	1 : 1174	1 : 17
1904	33,763,434	549,784	29,682	877	1 : 1138	1 : 18
1905	34,152,977	520,031	30,221	885	1 : 1131	1 : 17

Throughout the World (1915), traces the rise of the disease in American and foreign cities from 1881 to 1914; notice that the cancer rates were higher in the then more highly industrialized cities of Europe.

Today cancer is the second leading cause of death: 32% of Americans, or approximately 1 in 3, can expect to contract this disease sometime during their lives; 1 out of 4 men and 1 out of 5 women will die of it. In 1971 President Richard M. Nixon declared an all-out war on cancer in the United States. At the time, the death toll from the disease was 337,000. In 1986, 15 years after the "war" started, there were 472,000 deaths from cancer, an astounding 40% increase, despite the billions spent on research and treatment. Noteworthy in this context is the continued rarity of cancer among the primitive tribes of the world.

With this poor record of trying to solve the cancer problem by treatment, the time seems long overdue to devote our efforts to cancer prevention.

TABLE 16.10
Comparative Cancer Mortality in American and Foreign Cities, 1881–1914

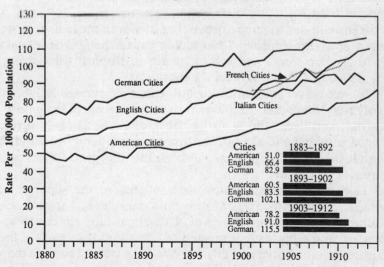

It is virtually impossible to avoid all the proven or suspected carcinogens that exist today. Such factors as air pollution and nuclear radiation are so ubiquitous that they are beyond the individual's control. We have no choice but to accept the fact that our bodies will be exposed to unwanted X factors from these sources. We can, however, choose not to subject ourselves to additional xenobiotics.

The free-radical action of cigarette smoke is a proven cause of cancer (and of coronary heart disease), and smoking is certainly within our control. Smokers are 10 times more likely to get cancer than nonsmokers. Smoking is the cause of 30% of all cancer deaths. The risk increases with the number of cigarettes smoked, the length of time one has been smoking, and how deeply the smoke is inhaled. However, lung cancer is not the only risk associated with the use of tobacco products (including chewing and dipping as well as smoking); cancers of the mouth, larynx, esophagus, pancreas, kidneys, and bladder may also be caused by tobacco use.

Chlorinated water as a cause of cancer has not been as highly publicized as smoking, but it has been shown to increase the risk of bladder, prostate, and other cancers. There is no evidence as yet that chloramine, an ammoniated chlorine compound used for water disinfection, causes cancer, but studies currently under way may prove that it is even more dangerous than ordinary chlorine. Chloramine causes hemolytic anemia in dialysis patients and instant death in tropical fish—strong evidence that it is a powerful xenobiotic.

In excessive amounts, alcohol contains enough solvents, polymers, and other carcinogens to cause cancers of the mouth, throat, and liver, and is also a cause of cirrhosis of the liver. Since smoking and alcohol consumption frequently go hand in hand, their toxic effects may also go hand in hand to increase the risk of cancer.

Talc, which is sometimes used to prevent the spoilage of polished white rice, was at one time considered harmless for human consumption. However, it is chemically similar to asbestos, and asbestos is suspected of contributing cancers of the gastrointestinal tract and ovaries. Washing treated rice, no matter how thoroughly, will not get rid of the talc. In recent years the dangers of talc have been recognized, and not all manufacturers continue to use it. But brown rice is healthier anyway, and the obvious alternative. Talc may also be found in salami, peanuts, and chewing gum. Shockingly, asbestos is still used by some bottlers to filter beer, wine, and soft drinks, and these beverages may contain asbestos fibers.

Polycyclic aromatic hydrocarbons are derived from coal tar. These proven carcinogens are found in polluted air and are ingested with such foods as smoked fish, barbecued meat, and coffee. Most synthetic food colorings are also polycyclic aromatic hydrocarbons.

Coffee, betel nuts, horseradish, tobacco, and various medicinal herbs are natural pesticides. Humans discovered the stimulating effects of these plants and used them for food, beverages, and smoking. Only later did we discover that they were associated with the development of many types of cancer.

Aflatoxins are produced naturally by the mold *Aspergillus flavus* on such foods as peanuts and grains under conditions of extreme heat and moisture. Along with hepatitis B virus, aflatoxins are suspected causes of liver cancer in the tropics. Under ordinary circumstances the grains and peanuts grown in the United States are not subject to such weather extremes, so this is not an American problem unless we import these foods from central Africa, Thailand, and other tropical areas. Importation of foods containing aflatoxins, however, is minimized by FDA inspection programs.

Nitrosamines are known carcinogenic compounds found in many foods. These cancer-causing substances can also be manufactured in our gastrointestinal tracts from nitrites or nitrates, which are found in some drinking water and used as preservatives in various foods, particularly delicatessen meats (salami, pastrami, corned beef, hotdogs, bacon, and others).

A diet high in protein, fats, and/or cholesterol (e.g., a diet of animal foods or excessive amounts of vegetable oils) has been correlated with cancer. The carcinogenic pattern occurs in the industrial world but does not appear in primitive meat-eating societies. (See Tables 16.11–14.)

TABLE 16.11
Animal Protein Intake and Intestinal Cancer

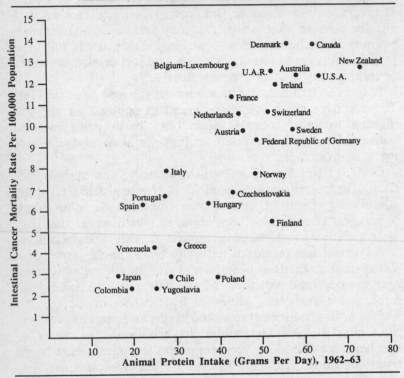

Source: O. Gregor, R. Toman, and F. Prušová, "Gastrointestinal Cancer and Nutrition," *Gut* 10 (1969):1031–34. Reprinted by permission of the author from Keith Akers, *A Vegetarian Sourcebook* (New York: G. P. Putnam's Sons, 1983).

The list of carcinogens that we consume is extensive. Carcinogens are xenobiotics, and they probably cause other diseases besides cancer. The foods and antioxidants recommended in the ten-point low-toxin diet are important in combating free-radical activity, cancer, and the other diseases of civilization. Exercise, relaxation, and stress reduction may also combat cancer. Studies have shown that exercise and stress reduction can

TABLE 16.12
Dietary Fat and Prostate Cancer

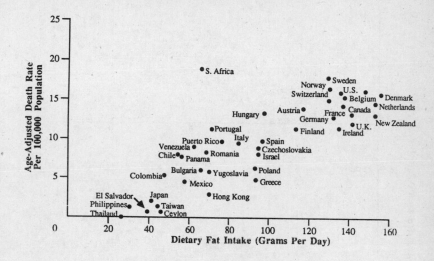

TABLE 16.13
Dietary Fat and Breast Cancer

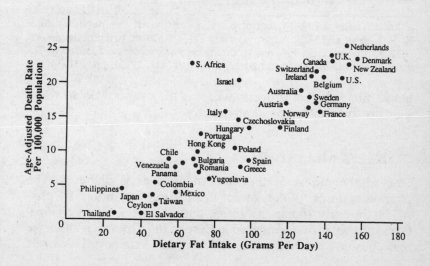

TABLE 16.14
Dietary Fat and Colon Cancer

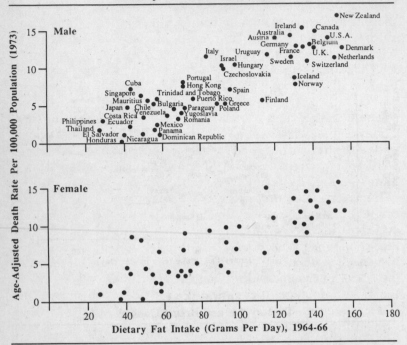

Note: Tables 16.12–14 reprinted by permission of the publisher from K. K. Carroll and H. T. Khor, "Dietary Fat in Relation to Tumorigenesis," in K. K. Carroll, ed., *Progress in Biological Pharmacology: Lipids and Tumors*, vol. 10 (New York and Basel: S. Karger AG, 1975), pp. 331–34.

help animals reject cancers, whereas they cannot do so when they are subject to constant stress.

DIABETES MELLITUS

One of the oldest of the modern diseases is diabetes. The phenomenon of "honey-sweet urine" was first noted by physicians in early Indian, Chinese, and Middle Eastern civilizations. Diabetes first appeared in Europe in 1673, when affluence and obesity were beginning to emerge in England. However, it remained a rare disease even as late as the nineteenth century. Sir William Osler, in his medical textbook *The Principles and Practice of Medicine* (1895), mentions this fact. At the time, there were 9 people with diabetes for every 100,000 people in Europe. Diabetes was even rarer in the United States, where the figure was 2 for every 100,000. Osler also says that out of 35,000 patients treated at the Johns Hopkins Hospital and Dispensary in Baltimore, Maryland, there were only 10 cases of diabetes.

These figures stand in sharp contrast to present statistics. Today diabetes and its complications are the third leading cause of death, after coronary heart disease and cancer. In 1985 there were approximately 12,000,000 known diabetics in the United States—about 1 in every 20 Americans.

It has long been thought that diabetes probably has a genetic origin, but there is reason to believe that this view is not entirely accurate. Surveys of the Bantu and other African tribes indicate that though rural tribespeople do not develop diabetes, there is a rising incidence of the disease among those who move to industrial towns and cities. Diabetic aborigines who return to the bush, abandon their modern lifestyles, and resume the hunter-gatherer existence of their ancestors experience a reversal of their diabetes.

ALZHEIMER'S DISEASE

Alzheimer's disease is a new ailment that was first described by a German physician, Alois Alzheimer. In 1907 he told a medical group about a 51-year-old patient who suffered memory loss and disorientation, gradually developed severe dementia, and died.

It has been estimated that Alzheimer's disease is the fourth most prevalent cause of death in the United States, though perhaps because of its rarity until recently, vital statistics tables do not include Alzheimer's (which is not necessarily the same as senility). Thus, it is difficult to get an accurate idea of its extent or to trace its pattern of emergence. However, most people involved with the study or treatment of Alzheimer's consider it more common today than it was 30, 40, or more years ago. Certainly, because of recent publicity, both doctors and the general public are more aware of its existence. Surveys conducted in Northern Europe suggest that as much as 58% of the over-65 population has Alzheimer's; about 3 million Americans suffer from the disease. From near obscurity, it has become a very common disorder indeed.

Alzheimer's disease has recently been linked to aluminum, high levels of which have been found in the brains of victims. Aluminum ions interfere with choline transport and thereby contribute to acetylcholine deficiency, a frequent characteristic of this disorder.

Aluminum did not become readily available until the late nineteenth century, when electric power made its production feasible for the first time. Thereafter, it appeared everywhere— in baking powder, antacids, buffered aspirin, tea, tap water, fabric softeners, antiperspirants, and processed cheeses. In addition, aluminum is found in asphalt, cooking utensils, wrapping foil, and cans. Dr. Alzheimer's discovery occurred approximately 20 years after aluminum was introduced and began to be widely used. Like coronary artery disease, Alzheimer's is unknown among primitive people living in jungles, rain forests, or other "undeveloped," nonindustrialized areas of the world (with rare exceptions, discussed below in the section dealing with amyotrophic lateral sclerosis).

Scientists have tried for years to determine whether there is a hereditary basis for Alzheimer's. One recent breakthrough that seemed to provide an answer was the discovery in Alzheimer's patients of reduplicated genes on chromosome 21 (as also occurs in Down's syndrome). Based on these genetic markers and the additional isolation of an unusual protein called amyloid B, a possible diagnostic test for Alzheimer's was proposed. Further investigation, however, could not confirm the original research, thus compounding the confusion over the possible role of heredity. But even if heredity is ultimately implicated, it does not follow that all bearers of reduplicated genes or other genetic traits will necessarily develop the disease. The role played by environmental xenobiotics is still poorly understood by scientists but is undoubtedly very important. Even if you have a family history of Alzheimer's, you can still exercise some control over your fate by following the principles outlined in this book.

MULTIPLE SCLEROSIS

Multiple sclerosis (MS) is a neurological disorder that usually afflicts young people. The earliest known case is recorded in the diary of Augustus d'Este, dating from the early nineteenth century. D'Este suffered his first attack at the age of 28; as seems typical of MS, he then experienced what appeared to be complete remission. He described subsequent attacks with such accuracy that his disease cannot be possibly mistaken for anything else.

Multiple sclerosis was recognized and defined through a series of studies performed by various European doctors from 1835 to 1872, but it still seemed to be relatively rare. There were no known reports of MS in the United States at the time. By the beginning of the twentieth century, the disease seemed more prevalent and began to be seen by more physicians. By the end of World War I, it was widespread in the United States.

At present, it is estimated that there are 250,000 victims of multiple sclerosis in the United States (58 cases for every 100,000 people), with an additional 250,000 sufferers of related disorders. Medical scientists believe that MS may have a genetic origin, since histocompatability locus antigens (HLA) studies show a higher incidence among people with similar chromosome makeups. However, clusters of MS cases do occur among strangers living within specific geographical areas, suggesting that there may be an environmental factor involved.

Although multiple sclerosis is not nearly as widespread as coronary heart disease, Parkinson's disease, cancer, or Alzheimer's disease, there is mounting evidence that the incidence of the disorder is rising in the United States, Canada, Europe, and Australia. Moreover, multiple sclerosis is nonexistent among such primitive peoples as the Eskimos and Bantus. This would indicate a genetic basis for multiple sclerosis were it not for the presence of the disease in Bantus and other tribespeople who leave their rural enclaves for industrialized areas.

AMYOTROPHIC LATERAL SCLEROSIS

The neurologic condition amyotrophic lateral sclerosis (ALS), also known as Lou Gehrig's disease, is a devastating one that strikes people down in the prime of life. It is fatal, usually within five years of onset. ALS was first described in 1869 by the famous French neurologist Jean Martin Charcot. There are no statistics on the number of people throughout the world who suffer from it, but it can strike anyone, regardless of race, sex, age, or economic status. However, it is not believed to be as prevalent as multiple sclerosis or Alzheimer's disease.

ALS takes its more common name from the New York Yankees baseball player Lou Gehrig, whose career was cut short by the disease. Other well-known victims include actor David Niven and ex-senator Jacob Javits of New York.

Amyotrophic lateral sclerosis is generally unknown in non-industrialized, primitive areas of the world. However, a high incidence of ALS, along with Parkinson's disease and an Alzheimer's-type dementia, has been found among tribesmen in western New Guinea and among the Chamorros of Guam and Rota. Initially scientists suspected that either a slow virus or hereditary factors produced these diseases. A slow virus was known to cause a neurologic disorder similar to ALS in northeastern New Guinea (the natives called it *kuru*, which is identical to the industrial nations' Creutzfeldt-Jakob disease and comparable to a neurologic disorder of sheep, scrapie); heredity was suspected because the ALS, Parkinsonism, and Alzheimer's of Guam was much more common in some families than in others.

What the early scientists overlooked was the possibility of an environmental factor affecting family members living together and eating the same foods. New research, though, showed convincing evidence that the causative agent was a naturally occurring, slowly toxic xenobiotic, the amino acid beta-N-methylamino-L-alanine (BMAA), found in the seeds of the cycad plants, notably the false sago palm. A common practice on Guam was to grind the cycad seeds and use the flour in a variety of foods or native medicines. The seeds were not originally suspect since the neurologic disorders often took years to develop, frequently emerging long after the cycad had been discontinued from the diet. The diseases began to increase in frequency after World War II, reflecting the increased use of cycad during wartime food shortages. Soon after the war the islanders' lifestyle changed; cycad seeds were no longer used as food, and within two or three decades the diseases began to disappear.

Closer to home are the cases of three former players on the 1964 San Francisco 49ers football team who were stricken with ALS; two died and one remains seriously ill. Two playing fields used by the team were believed to have been treated with Milorganite, a human waste fertilizer possibly contaminated with cadmium and other heavy metals. In addition, two employees of the Milwaukee Metropolitan Sewerage District (the producer and distributor of the fertilizer) who worked with or around Milorganite also died of ALS. Though several scientists object vigorously to this purported linkage, others insist that studies are needed to prove or disprove it. In the interim, it is prudent to be aware of the possible association and avoid unnecessary exposure to the fertilizer.

Also suggestive is a severe form of Parkinson's disease that develops rapidly in drug users, scientists, and experimental animals exposed to the synthetically produced "designer drug" MPTP—additional support for the link between xenobiotics and neurological disorders.

DIVERTICULOSIS

Medical texts and surgical papers first took note of diverticulosis in 1911. At present, it is fairly widespread in the Western world, with over 55% of all adults over the age of 60 estimated to be afflicted. It still remains relatively rare among primitive people. In a recent autopsy study among the Bantu people of

South Africa, only 5 cases were found out of 3,000 autopsies performed.

Diverticulosis is characterized by the development of little pockets or pouches in the lining of the large intestine (colon). It can be symptom-free, but if food gets stuck in the pouches, it can cause severe pain. Further, inflammation can develop, leading to diverticulitis, which is potentially very serious. Cancer of the colon is more common in patients with diverticulosis and diverticulitis.

It is commonly believed that diverticulosis is caused by a lack of fiber or roughage in the diet. This is at least partially correct. However, the disease does not develop in primitive meat-eating peoples who have virtually no fiber in their diet.

APPENDICITIS

Surprisingly, there is some evidence that appendicitis is also a new disease. In the early nineteenth century, it was extremely rare, if not actually nonexistent, in the United States, Canada, and Europe. Abscesses near the cecum were discovered in the middle of the nineteenth century when doctors began to perform autopsies on a regular basis. But even after this discovery, there were relatively few cases. Between 1840 and 1860, only 33 cases were noted in all of Europe and North America. The identification of these abscesses as caused by perforation of the appendix, and the surgical treatment of appendicitis, were suggested by Dr. Reginald Heber Fitz of Boston in 1886.

At the turn of the century, when diagnosis and surgical treatment were finally well established, 5 to 10 cases of appendicitis were treated at the Radcliffe Infirmary in Oxford, England. In 1970 that same establishment had only a moderately greater number of total patients, yet it treated 500 cases of appendicitis. The statistics for the United States are similar. However, the disease appears to have spread first in the white population and later among blacks. As late as 1920, physicians practicing in the "black belt" of Alabama had never seen a case of appendicitis.

SARCOIDOSIS

The disease known as sarcoidosis is an interesting phenomenon. It most often strikes young adults, and while it can afflict all ethnic groups, the white populations of Norway, Sweden, and Ireland seem to show a particularly high rate of incidence. In the United States, however, it is the black population that is most affected by sarcoidosis, and their rate of incidence is the highest in the world. This is particularly noteworthy since the disease is totally unheard of among the rural blacks of Africa.

A multisystem disease, sarcoidosis was first described by the Danish physician Caesar Boeck in 1899. It was originally called Boeck's sarkoid because it causes granuloma, or lesions, that resemble sarcoma, a kind of cancer, when they appear on the skin. There is no known cure for the disease, and because the possible effects of xenobiotics have been ignored, attempts to determine the cause have been futile.

SYSTEMIC LUPUS ERYTHEMATOSUS

Many people have never heard of systemic lupus erythematosus (SLE). It is a disease that afflicts young women more often than men; 9 out of 10 lupus victims are women, usually between 15 and 30 years of age. Lupus is a chronic systemic inflammatory condition involving the immune system. It is in the same family of diseases as rheumatoid arthritis.

The first description of lupus was presented in 1828 by the French physician Laurent Théodore Biett. Forty-five years later Moritz Kaposi, an Austrian physician, showed that lupus not only caused skin rashes but could also affect the internal organs. In 1890 Sir William Osler found that lupus could affect the body internally without causing skin damage.

The cause of lupus is obscure, but it is yet another condition where the evidence seems to point to xenobiotics; it is estimated that 10% of lupus patients develop the disease after taking various toxic drugs. The most common villain is procainamide (Pronestyl), which is used to treat irregular heartbeat.

Today, the incidence of lupus is rising (though again, the

disease is strikingly absent among primitive peoples). Much of the increase is real and not just a function of improved diagnosis. At present, 1 in 400 American women have lupus. It is more common than Lou Gehrig's disease, hemophilia, multiple sclerosis, leukemia, cystic fibrosis, or muscular dystrophy. One million Americans are believed to have SLE, with 55,000 new cases diagnosed each year.

HYPERACTIVITY AND LEARNING DISABILITY

Hyperactivity and learning disorders (also known as attention deficit disorder) may be new conditions. The first known article concerning reading disabilities in otherwise intelligent children was published in 1896 by British ophthalmologist N. Pringle Morgan.

There are strong indications that xenobiotics have played an important part in at least some, if not many, cases of hyperactivity. Dr. Ben Feingold developed the Kaiser-Permanente Diet in the 1950s, after he observed that many of the hyperactive children under his care improved when he withheld salicylates and synthetic colorings and flavorings from their diets. Later he discovered that some children also reacted adversely to BHT (butylated hydroxytoluene), a phenolic preservative found in many foods.

Classic nutrition experiments were conducted in India at the turn of the century by Sir Robert McCarrison, who was a medical officer in the Royal Indian Army. These experiments demonstrated the strong connection between diet, health, and behavior. Sir Robert was assigned to practice medicine in the town of Gilgit, high in the Himalayas, in the extreme north of Kashmir. Sixty miles north of Gilgit lived the Hunza, an extremely long-lived people whom he rarely visited because they were rarely ill. In all respects their health was excellent, unlike that of the peoples of the surrounding areas, who suffered from a variety of illnesses.

Sir Robert began to wonder about the reasons for this. The climate and sanitary conditions were the same for the Hunza and the other peoples of the area. The only apparent difference seemed to be diets. He decided to experiment on rats, feeding groups of them the various tribal diets. The Hunza rats were very healthy; they grew rapidly, had pleasant dispositions, were never ill, mated with enthusiasm, and had healthy offspring, unlike the other rats, which were snarling and vicious. At 27 months, the rats were killed and autopsies performed. The Hunza rats were indeed free of disease. The other rats had diseases that matched those of the people whose diets they had been fed, diseases involving all of the body's organs.

More recently, authorities at the U.S. Naval Correctional Center in Seattle found that removing white bread and refined sugar from the diet of inmates reduced the incidence of violent behavior. A year later, authorities at the Tehama County Juvenile Hall in Red Bluff, California, substituted honey for sugar and eliminated ham, packaged foods, preservatives, artificial colorings, and flavor enhancers from the diet of children under their supervision, with similar positive results.

When Pam Crook, staff nutritionist at the Canyon Verde School in Hermosa Beach, California, was given the unenviable task of trying to solve the behavior problem of a hyperactive student, she persuaded the student and the student's mother to avoid junk foods, high nitrates, salt, sugar, and foods with additives and preservatives. The results were dramatic; the student had a complete turnabout in personality.

After this success, Pam decided to conduct a rat experiment for the benefit of the students, to show them how diet can affect behavior and health. Their project ran for three months and utilized four families of small rats, which lived in four separate cages. Each group was fed a consistent diet of laboratory pellets, another ingredient, and a beverage. The first cage was fed pellets and clean water; the second cage, pellets, commercially prepared hotdogs, and tap water; the third cage, pellets, sugar-coated breakfast cereal, and fruit punch; and the fourth cage, pellets, sugared donuts, and cola.

The results were quite distinct. The rats in the first cage remained alert, curious, and social. Those in the second cage, the ones fed hotdogs, became extremely violent, exhibiting belligerent behavior and uncoordinated, jerky movements. The rats in the third cage, who had been given sugared cereal and fruit punch, were hyperactive, with constant frenetic movements. The rats on the diet of sugared donuts and cola appeared to be autistic. They sat in their cage isolated from each other, rocking back and forth, with jerky, erratic movements.

Although we need more research to prove that xenobiotics are implicated in attention deficit disorder and antisocial behavior, the available evidence is extremely suggestive. I believe that many afflicted individuals would be helped by diet modification and avoidance of such currently prescribed mind-altering drugs as Ritalin.

ALLERGIES

Although milk allergy was described by Hippocrates around 400 B.C., allergies remained rare throughout history. The first American report of milk allergy appeared in 1916, the first British report in 1958.

There is evidence that allergies are becoming more widespread and more serious in the developed world. For example, in the late 1890s Sir William Osler reported that asthma was a rare, mild condition in both children and adults; it almost never caused death. Recently asthma has become more common, in some cases decidedly serious, and is causing an increasing number of fatalities each year, despite rapid medical advances. Nasal allergy, too, was rare in Britain and the United States 50 years ago. Due to a scarcity of subjects, allergy research projects were few and far between. Today, the opposite is true. Are allergic disorders a product of industrial society? I believe that there may be a link. In many primitive areas of the world, asthma and other allergies remain rare.

Desensitization injections, a standard form of treatment, are effective for respiratory allergies but not for allergies to foods and toxins in the environment. Those who have allergic reactions to wheat, grains, milk, chocolate, or citrus fruit should eliminate these foods from their diet, but under supervision.

By strengthening the immune system and minimizing toxin intake, the ten-point program should help you avoid allergies.

Chapter 17

Primitive Versus Industrialized Societies

To understand the toll the Industrial Revolution has taken on our environment and our health, it will be useful to look at some of the remaining nonindustrialized societies and to compare their lifestyle to ours.

PRIMITIVE SOCIETIES

The Tarahumara Indians

Approximately 300 miles south of El Paso, Texas, in the Mexican state of Chihuahua, close to the Copper Canyon, lives a group of Indians called the Tarahumaras. They are almost totally isolated from the surrounding Mexican communities. They have escaped the encroachments of industrial society because their land is extremely mountainous, with many high sierras, canyons, and wilderness areas.

The Tarahumaras are natural athletes; they can run for hundreds of miles. In earlier days, when game was more plentiful, they were known to run after deer until the deer gave up from exhaustion. They built up their stamina by running up and down steep mountainous paths, at times carrying animals whose weight exceeded their own.

Running remains important to them as part of their religious rituals and as sport. In September and October they hold something akin to a track meet, engaging in a sport they call *rarajipari*. Teams from the various villages gather to kick a wooden ball over steep precipices for many miles—sometimes more than 150 miles. The villagers place bets on their teams; they all drink *sohuiki*, an alcoholic beverage made from cornmeal, until they are completely drunk, and then the game commences. It continues until the ball is kicked to its predetermined site—or until one of the drunken players falls from a precipice—in which case the team loses. It is not a sport for the weak-hearted.

With a few exceptions (one of them the playing of violins, introduced to them by the Spanish conquistadores), the Tarahumaras have preserved their ancient culture and lifestyle. They continue to live in caves, as they have for centuries, though some of their wealthier brethren have managed to move into crude huts. However, neither group has the benefit of electricity or indoor plumbing. Their water is not disinfected or piped in but comes directly from nearby streams and springs.

The Tarahumara diet consists of homegrown corn, cornmeal, squash, beans, and any local wild fruit and berries they manage to find. Although they do not eat meat regularly, they are not vegetarians, as is commonly believed. They eat deer, other game animals, and field mice when they can find them and catch them. Some of the more prosperous people raise goats and sheep, primarily for wool, milk, and cheese, but they occasionally slaughter animals for the feasts associated with their religious rituals.

Among the Tarahumaras, the major causes of death, besides old age, are accidents (mostly falling while under the influence of *sohuiki*), infant diarrhea, and other infections. They do not show any sign of high blood pressure, nor do they suffer from cancer or coronary heart disease. Their serum cholesterol levels average 125 mg/dl, which is extremely low.

The Eskimos

As most people are aware, the Eskimos are a group of Indian tribes living in the farthest northern stretches of habitable land in Alaska, Siberia, Canada, and Greenland. Although their lifestyle is changing, a good deal is known about their immediate past. The Eskimos were hunters and fishermen. A large part of their food supply was made up of fresh or frozen fish, fowl, eggs, and meat, all raw. For this reason their neighbors to the south, the Naskapi, called them Askimowet, or "raw meat eaters," a term that became "Eskimo." The Canadian Eskimos prefer the name Inuit, which means "men."

The Inuit dwelling was the well-known igloo, made by carving blocks of frozen snow and placing them together to form a wind barrier and shelter from the elements during the frigid months of the year. They used animal skins for warmth within the igloo and sewed them together to make tents during the brief months of warm weather.

The Eskimos had plenty of the purest drinking water possible, obtained from snow, freshly distilled by nature, with no minerals and no disinfecting compounds. Their diet bore no resemblance to our concept of a well-balanced diet composed of the four basic food groups. It consisted almost entirely of meat and fish, including bear, caribou, fox, musk-ox, seal, walrus, salmon, and whale. Frequently they would feast on pure blubber, eating pounds of it at one sitting. Birds and eggs were also a part of their menu when available. They had virtually no vegetables, cereal grains, bran, or fruit.

The traditional Eskimo diet was extremely high in saturated and unsaturated fats, cholesterol, and protein. Except for bits of moss and occasional berries and seaweed, it was totally devoid of roughage, greens, grains, fiber, and carbohydrates. It is the kind of diet that we in industrial society would consider certain to produce coronary heart disease. Fat consumption was in the range of 60% to 80% of total calories—far above the American average of 40%, and frightening when compared to the 30% or less recommended by the American Heart Association.

However, the incidence of coronary artery disease among traditional Eskimos is low.[1] One explanation is that they consume large amounts of essential fatty acids. Cold-water fish such as salmon contain large amounts of unsaturated fatty acids, including the omega-3 group, eicosapentaenoic acid (EPA) and docosahexaenoic acid (DHA). Unlike another fatty acid, arachidonic acid (AA), EPA does not induce platelets in the blood to stick or to form clots; as a result, Eskimos have a bleeding tendency. Other blood factors seem to be diet-related: Eskimos have low levels of cholesterol, triglycerides, LDLs, and VLDLs, and high levels of HDLs. These factors, along with the bleeding tendency, appear to contribute to the low incidence of coronary heart disease in Eskimos.

Arctic animals live in a pollution-free environment and may produce fatty tissue uncontaminated by toxins; fish caught in polluted waters may not offer the same benefits as the fish caught by the Eskimos.

Contrary to popular belief, Eskimos tended to be lean. It was the roundness of their faces, combined with their bulky caribou clothing, that led to the myth of their obesity, which has been disproven by skinfold measurements and height/weight calculations. Today, however, the myth is becoming reality. With the progressive Westernization of their diet, and a more sedentary lifestyle in their settlements, Eskimo men and women are becoming obese. In stores, they are buying refined bleached sugar, bleached white flour, salt, eggs, canned goods, meats, and cakes. Smoking has become an accepted ritual. They have forsaken the igloo for houses in Hudson's Bay Trading Company, government-sponsored settlements, or other permanent dwellings. The only difference between these houses and similar models in the United States and Canada is that they cannot have indoor plumbing because of the frigid temperatures; san-

---

[1]Paradoxically, despite the rarity of coronary artery disease in non-Westernized Eskimos of our own era, there is autopsy evidence of atherosclerosis in a 1,600-year-old Eskimo mummy. Atherosclerosis, or hardening of the arteries, is not synonymous with coronary artery disease. It has existed for centuries, whereas heart attacks are largely a phenomenon of the last 100 years.

itized water is trucked in, and sewage is trucked out. Their clean air is also changing; the Arctic skies in recent years have been clouded by a hazy orange-brown mist believed to emanate from the industrial discharge of factories in Siberia.

In their own habitat, before the intrusion of modernity, Eskimos did not develop multiple sclerosis, diabetes, lupus, cancer, and other of the "degenerative" diseases.[2] Now their health patterns are changing along with their lifestyle. Before the move to the settlements, the three leading causes of death among Eskimos were accidents, tuberculosis, and pneumonia—the latter two a legacy of European explorers. They were also susceptible to simple infections such as colds and to childhood diseases such as chicken pox and measles. Now obesity and diabetes are relatively common, and cancers of the breast, esophagus, colon, nasal passages, salivary gland, and stomach are no longer rare.

The Masai and the Samburu

The Masai of the East African Rift Valley between Kenya and Tanzania are a nomadic people whose lives revolved around the herding and raising of cattle. Since they were pastoral, their dwellings were never permanent, and they did not have any of the amenities of civilization, such as electricity, insulation, or indoor plumbing. The water they drank was never disinfected with chlorine, nor was it transported in pipes of any sort. Their water came from springs or streams, often brackish and highly mineralized, and it tended to be rather scarce.

The Masai are another of the cultures that prove to be an embarrassment when it comes to the theory linking a high-fat diet with coronary heart disease. The Masai tend to be tall and slender, with no weight problems. The extent of their physical activity is not accurately known, but they are active while tend-

[2]One exception is osteoporosis, or softening of the bones, a very common disorder among menopausal women in industrial society. Historic data show that Eskimos did develop osteoporosis, probably because of their high-meat diet; meat sources of food contain large amounts of phosphorus and phosphoric acid, which remove calcium from bones.

ing their herds. Because of the hot climate, they nap for several hours during the sweltering afternoon. The usual daily meal of the *murani*, the adult males, is milk, both fresh and curdled, with sporadic feasts of meat. In the dry season, when the milk supply diminishes, cow's blood is mixed with the milk. The amount taken from each cow is limited to about a quart a year, so their consumption of blood is not extremely high. The *murani* diet is about 66% fat. The other members of the tribe— the women, children, and elders of both sexes—eat vegetables and legumes in addition to milk and meat.

Despite the high fat content of their diet, the *murani* have serum cholesterol levels in the 120 mg/dl range, well below the American average of 220 mg/dl. (See Table 17.1; the Samburu diet is described below; Pygmies are omnivorous, eating both animal and vegetable foods.)

Nathan Pritikin, in *The Pritikin Promise*, used the Masai as an example demonstrating the correlation between atherosclerosis and a diet very high in saturated fat. However, there have been several conflicting medical studies of atherosclerosis in the Masai. Although more recent studies confirm Pritikin's contention, earlier studies showed low blood cholesterol levels and an absence of atherosclerosis. It seems that as progress encroaches on the Masai, their tendency to develop atherosclerosis is increasing.

The Samburu tribes are closely related to the Masai and live nearby. Though their lifestyle is somewhat different, their story is similar. The Samburu raise camels, drink camel's milk and blood, and eat camel meat. Yet until recently, just like the Masai, they appear to have been free of the diseases of industrial society.

There has been no adequate explanation for the excellent health history of the Masai and Samburu. Some have suggested favorable heredity, but the Masai are considered to be of mixed ancestry because for 10,000 years their main occupation was stealing cattle and women from other tribes. Lack of stress also seems an unlikely explanation, since the Masai have stressful problems—among them wars—just as industrialized societies do.

TABLE 17.1
Cholesterol Levels in Diverse Cultures

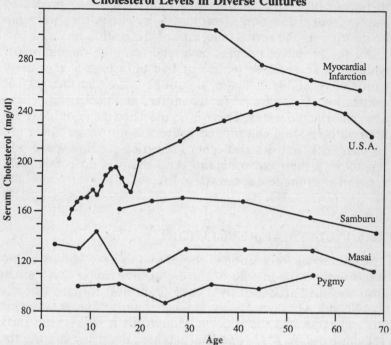

Source: U.S. Congress, Senate Select Committee on Nutrition and Human Needs, Hearings on Diet Related to Killer Diseases, vol. 3, *Red Meat* (March 24, 1977), p. 261.

Apart from exercise, there is only one factor that appears to account for the health of the Masai and Samburu—the one factor they have in common with the other healthy nonindustrialized cultures: relatively pure sources of food and water. They drank water that was not altered by chlorine and other disinfectants. Their cattle or camels had access to pure water and to forage untainted by industrial smoke or chlorocarbon and other pesticide residues.

Unfortunately, both the lifestyle and the health of the Masai and Samburu have undergone changes. They no longer live an isolated life, since the Kenyan government has attempted to move them to permanent settlements. They now have access to flour, sugar, confections, shortening, and other processed staples from civilization's factories. Very likely, they will soon be developing the serious diseases of civilization as well.

All the primitive peoples have one common denominator, whether they are meat eaters or live on a high-fiber, purely fruit and vegetable diet: they do not have the "advantages" of civilization. There are no food additives and packaging. There is no chlorinated water. There is no bleached sugar or bleached flour with residual chlorinated hydrocarbon toxins. There are no chemical factories and other industries pouring waste pollutants into their environment. And there is no comparable array of man-made diseases (see Table 17.2).

THE INDUSTRIAL REVOLUTION

The Industrial Revolution is considered to have had its beginnings around the middle of the eighteenth century. Although iron was first produced by a coke-fired blast furnace as early as 1709, by Abraham Darby, it was not until 50 years later that his son improved the process enough for it to be used on a regular basis. In 1733 John Kay patented the flying shuttle, the first of a series of inventions that would revolutionize the textile industry. In 1764 James Hargreaves invented the spinning jenny, and in 1769 Richard Arkwright invented the spinning machine. The power loom was developed by Edmund Cartwright in 1785, and in 1793 Eli Whitney invented the cotton gin.

Most people think of the textile industry when they think of the origins of the Industrial Revolution, but advances were taking place in other industries as well. In the mid-eighteenth century England and Europe began producing Meissen, Sèvres, Villeroy & Boch, Haviland, Wedgwood, and other fine china and pottery.

Industry became so important during this time that the first of many expositions of the industrial arts was held in Paris in 1763. However, change was gradual until the latter part of the eighteenth century, after James Watt finally managed to put his steam engine to practical use in 1776 and C. L. Berthollet

TABLE 17.2
Diseases Common in Industrialized Societies but Not Found Among Primitive Peoples

Gastrointestinal	
diverticulosis	ulcerative colitis
diverticulitis	irritable colon
appendicitis	cancer of the colon and
constipation	rectum
gallstones	polyps
regional ileitis (Crohn's	hemorrhoids
disease)	duodenal ulcers[a]

Cardiovascular and Metabolic	
coronary heart disease	peripheral artery disease of
angina	the legs
stroke	varicose veins
high blood pressure	deep vein thrombosis
diabetes mellitus	pulmonary embolism
obesity	kidney stones
gout	

Endocrine, Autoimmune, and Others	
thyrotoxicosis	pernicious anemia
myxedema	cancers of various types
Hashimoto's thyroiditis	Alzheimer's disease[c]
Addison's disease	tooth decay
systemic lupus erythematosus	allergies
hypoglycemia	eating disorders
rheumatoid arthritis	phobias
multiple sclerosis	panic reactions
amyotrophic lateral sclerosis[b]	suicide
Parkinson's disease[b]	childhood behavior disorders
senile osteoporosis	various mental disorders
Paget's disease	sarcoidosis
subacute combined	endometriosis
degeneration	premenstrual syndrome

[a]Rare in most primitive areas; a few cases found in tribal Africa.

[b]Unknown among primitive African blacks, but cases have been found among the Auyu and Jakai people of western New Guinea.

[c]A high incidence of Alzheimer's, along with ALS and Parkinson's disease, was found in Guam.

introduced the process of chemical bleaching with *eau de Javelle* chlorine in 1785. From then on, mechanical inventions and chemical discoveries proliferated so rapidly that they all but defy enumeration (see Appendix A for an extensive but by no means complete list).

People's lives began to alter, perhaps imperceptibly at first. Factories were built, providing many new jobs but also sending chemical wastes into rivers and streams. Workers, many of them children, were brought into daily contact with chemicals and substances they had never heard of before, much less touched. New products began to appear in the marketplace, in greater quantities and to a greater standard of uniformity than had been possible by handcrafting. But perhaps most important of all, people began to believe that almost anything— from eating utensils to food to fabrics to transportation—could be improved scientifically by mechanical inventions or chemical alteration.

It was in the nineteenth century that the Industrial Revolution began to make enormous strides, and with them enormous changes in lifestyle and habits. The steam engine brought rapid development of railroad and steamship travel. Food was first preserved in glass containers in 1810; the following year the tin can was invented, and in 1819 the commercial canning industry was born. Gaslights were introduced in the early 1800s and made way for the electric light before the century was out.

Machinery was developed for all sorts of manufacture in the nineteenth century, from lead pencils and paper to glass and steel. Companies began drilling and piping oil as more and more uses were found for it. Meatpacking became a major industry with the development of the transportation network and the refrigerated car. Chemists produced matches, potash for fertilizer, photosensitive bromine, iodine, and silver salts for photography, pigments for paint from coal tar, zinc, lead, and copper, synthetic dyes for cloth and food. The number of new products and industries kept mounting as the century produced dynamite, dishes, flatware, soaps, and an entirely new commercial enterprise, pharmaceuticals.

With all this came an ever increasing need for water—to

create steam, to cool motors, as a raw material for the manufacturing process, or as a convenient medium for the disposal of chemical wastes and by-products. As people left the rural areas to take the new jobs provided by industry, the urban population soared, creating a tremendous demand for drinking, cooking, and cleaning water. Other factors such as the introduction of indoor plumbing, the establishment of public baths, and the threat of fire in large population centers intensified the need for municipal water supplies.

The cities were not prepared to meet these rapidly rising water requirements, even though most of the major cities by necessity were located on waterways. At first water mains were made from hollowed wooden logs, but these were not entirely satisfactory and were soon supplanted by lead pipes or by iron and steel pipes lined with coal tar to prevent rust. Further technological advances brought cement pipes lined with asbestos. Many of these same water delivery systems are still in use in some of the world's older cities.

But health was not a major concern in the development of these systems or of the other nineteenth-century technologies. The toxic properties of lead, asbestos, and coal tar were unknown then. The germ theory of disease had not yet been formulated, so raw sewage from homes and runoff from storm drains were permitted to flow back into the rivers to be recycled. The chemical industries, the cloth-dying companies, the felt hat manufacturers, and all the other new industries discharged their waste into the waterways or buried it underground. They also produced masses of chemical-laden smoke that became part of the air and eventually settled on buildings, water, land, and the crops and forage growing on the land.

All of these substances—wastes, chlorine compounds, etc.—became part of the food and water chain in industrial society. The timing of these events is important, because the proliferation of chemicals in the environment parallels the emergence of the man-made diseases.

As the century progressed, other changes were also taking place that would eventually add to the problem. The growth of the canning industry brought new ways of preserving food or altering it to make new and different food products. In the late 1860s and early 1870s the meatpacking industry began. To aid in distribution, the refrigerated railway car was developed. In 1886 a means was found for obtaining aluminum from bauxite, and this material became plentiful. In 1912 cellophane was invented; freezing of food became possible in 1917; and plastic packaging was introduced in 1945. It would be impossible to detail the extent to which foods began to be processed, altered, and adulterated during this time, but one example is a list of additives to coffee that James R. Mann read in the U.S. Congress in 1906: the colorings Scheele's green, iron oxide, yellow ochre, chrome yellow, burnt umber, Venetian red, turmeric, Prussian blue, and indigo, along with roasted peas, beans, wheat, rye, oats, chicory, brown bread, charcoal, red slate, bark, and date pits. That year, after a decade of public awareness of the problem, Congress passed the Pure Food and Drug Act. The federal government had been petitioned by all the state dairy and food departments for such a law eight years earlier, and similar laws had been passed in England in 1860 and 1872.

Another significant development of the late nineteenth and early twentieth centuries was the commercial production of cigarettes. Earlier, tobacco had been enjoyed principally by the wealthy, in the form of cigars, pipes, snuff, or chewing tobacco. The cigarette was developed in Europe and introduced to the United States in 1867. However, cigarette smoking was not widespread until much later, because the cigarettes were too strong to inhale and had to be rolled individually. In 1890 the flue-curing of tobacco made the leaf much milder, and by 1900 the mass production of cigarettes had reached the 4 billion mark. Prepackaged cigarettes were first marketed in 1914, and with promotions such as free packs for servicemen in World War I, the cigarette coupon, and testimonials from entertainment stars, the business boomed.

The internal-combustion engine was another step forward that has had unintended consequences. Early in the nineteenth century, attempts were being made to develop a practical internal-combustion engine and an appropriate fuel. Most flammable liquids were ineffective and inefficient. Kerosene, developed in 1854, came close to doing the job, but oil proved to be the ideal solution. The first oil well was drilled in the United States in 1859, and by 1865 the first pipeline had been built and the first company was organized to sell natural gas.

Around the same time, various inventors were developing the horseless carriage. The earliest models were steam-powered, but in 1885, in Germany, Daimler and Benz managed to produce the first automobiles using an internal-combustion engine. By 1892 similar cars were being made in the United States, and the same year Diesel produced a heavy oil machine. By the 1920s the roads of the industrialized world were filling with cars and trucks and buses, and airplanes were taking to the skies in sizable numbers. By 1939 the jet plane had been developed, as well as the helicopter. Soon automobile and airplane exhaust fumes were polluting the air of the world's major cities. By the 1960s smog had reached a crisis point throughout the world, and people began to talk about "acid rain."

At about the same time the oil industry had its beginnings, the first synthetic polymers were produced. (Polymers are very large molecules made of tens of thousands of small ones; they are virtually indestructible.) Nitrated cellulose was developed in 1862, and celluloid in 1870. Celluloid was used as a substitute for ivory and in a variety of products, from shirt collars to billiard balls to photographic and motion picture film. The first thermosetting plastic was Bakelite, developed in 1909, and the first synthetic fabric was rayon, invented in 1902. Nylon came along in 1934, and Acrilan, Orlon, Dynel, and Dacron in 1941. The first silicon polymers were developed in 1945.

In the 1930s chlorinated petrochemical compounds were developed for a wide variety of applications in industry and agriculture. After World War II chemical technology became even more sophisticated, flooding the market with a vast number of highly toxic new chemicals: DDT, dieldrin, methyliso-cynate, etc. In 1948, Dr. Paul Müller won a Nobel Prize in medicine and physiology for his development of DDT; 14 years later, Rachel Carson pointed out the disastrous price the environment was paying for Müller's brainchild in her dramatic warning, *Silent Spring*. With few exceptions, these synthetic chemicals are very stable, which means they remain in the environment for a long time. Some have already contaminated the groundwater and entered the food chain.

No one could have predicted the results of all the important discoveries and inventions of the past 200 years, nor could anyone have foreseen that toxic chemicals and waste products would enter the food chain and end up in the human body. But in retrospect we can see that the gradual development of the Industrial Revolution precisely parallels the rise of the deadly diseases of the twentieth century (see Table 17.3). The evidence is so overwhelming that we can safely reject the possibility of coincidence; it is clear that today's killer diseases are virtually all man-made.

TABLE 17.3
The Industrial Revolution and the Increase of Cancer

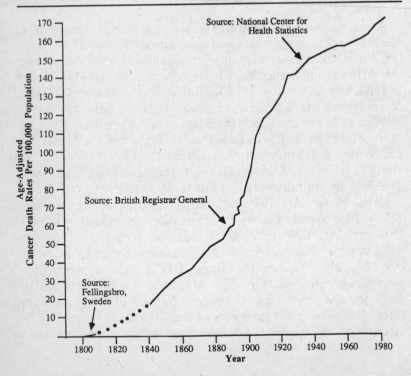

Chapter 18

The Causes of the Man-Made Diseases

For at least the last 200 years we have been eating, drinking, and breathing low-grade poisons from our environment in increasing amounts. These accumulate and interact within our bodies, causing chemical changes that may be responsible for many of our present-day diseases.

The industrial nations of the world are being overwhelmed by xenobiotics and free-radical producers: ultraviolet irradiation, nuclear irradiation, X-rays, microwave energy, acid rain, cosmic irradiation, radon emissions, singlet oxygen, and many others. Many of these we can do very little about. But there is also an ever increasing number of man-made substances with toxic potential. Heading the list are tobacco smoke and tars, dyes, industrial chemical wastes, the halogens (chlorine, bromine, fluorine, iodine), and halogenated hydrocarbons, followed by ozone and nitrous oxides in smog, gasoline, solvents, automobile and jet engine exhaust, petrochemicals, and pesticides, including organophosphates, methylisocyanate, aldicarb, dieldrin, and DDT. Then there are the synthetic food additives: colorings, flavorings, sweeteners, and preservatives.

SOME COMMON XENOBIOTICS

There are far too many common sources of contaminants to list them all. Every day we come into contact with engine exhaust, tobacco smoke, dyes, lead, cosmetics, petrochemical toners in copying machines, and hundreds of other toxins. What follows is a discussion of some of the more widespread industrial and chemical xenobiotics that appear to cause the greatest health hazards.

Chlorine

Chlorine has seemed an unlikely villain because of its value in disinfection. The addition of chlorine to drinking water beginning around 1900 was an important health breakthrough. Swiftly many of the dreaded water-borne diseases that had been major killers in previous centuries were virtually eliminated. However, some researchers suspect that chlorine was one of the factors responsible for the appearance of coronary artery disease around 1910.

Research done at the turn of the century indicated that chlorine was very effective against bacteria without causing any ill effects in humans. For this reason, people have always had confidence in the supposed purity of tap water. But this early research was concerned primarily with disinfection and chlorine's ability to fight coliform bacteria and disease-causing germs, not with the possible effects of prolonged use of chlorinated water.

There are now indications that many public water supplies are of questionable purity. Chlorine is not inert; it interacts with other chemicals and even simple organic matter such as algae, bacteria, and humic acid. These chemical reactions sometimes produce chloroform and other known carcinogens. In addition, industrial wastes, sewage, and agricultural runoff mix into our water supplies. The result is that there are literally thousands of toxic chemicals in the water, some of which also react with chlorine to create even more toxic compounds.

Known or Suspected Carcinogens
in Finished Drinking Water

vinyl chloride	aldrin
carbon tetrachloride	heptachlor epoxide
trichloroethylene	diphenylhydrazine
bis (2-chloroethyl) ether	benzene
dieldrin	benzo[a]pyrene
heptachlor	1,4-dioxane
chlordane	methyl iodide
DDT	DDE
beta-BHC	1,2-dichlorethane
alpha-BHC	1,1,2-trichlorethane
chlorinated biphenyls	simazine
chloroform	tetrachloroethylene
bromodichloromethane	acrilonitrile
chlordibromomethane	1-bromobutane
bromoform	fluoroform
bromodifluoromethane	trifluoroethylene

Compounds Formed in Drinking Water
by Reaction with Chlorine

chloroform	trichloroacetic acid
bromodichloromethane	trichloroacetaldehyde
chlorodibromomethane	chlorophenols
bromoform	chlorobenzenes
dichloromethanes	a-chloroketones
bis(2-chloroethyl)ether	chlorinated aromatic acid
trichloroethylene	chlorinated purines
tetrachloroethylene	chlorinated pyrimidines
1,1,1-trichloroethane	nonhalogenated compounds

Pesticides

DDT, lindane, dieldrin, dibromochloropropane (DBCP), and numerous other pesticides have been used for years in industrialized society. Pesticides were introduced to eliminate insects, rodents, mold, and other threats to agriculture and crops. DDT was also used in jungle areas to kill mosquitoes, which transmit malaria and other diseases.

DDT has now been shown to accumulate in the flesh of fish and birds, causing abnormally fragile eggshells and reproductive difficulties. The effects on humans are not quite so obvious. Unless one is exposed to a large amount of DDT all at once—enough to cause acute poisoning—the DDT is stored in body tissue and does its damage years later, in most cases not even suspected as the culprit.

The pesticide carbaryl (Sevin) can produce birth defects, cancer, and mutations. When workers involved in its production were tested, an extraordinary number had lowered sperm counts and damaged sperm, and were sterile to all intents. The EPA, however, has not banned Sevin, which is sprayed indiscriminately on crops throughout America. A highly toxic ingredient of Sevin, methylisocyanate, was involved in a serious industrial accident on December 4, 1984, in Bhopal, India; several thousand people died, and many more were left with permanent blindness and chronic respiratory illnesses.

Since the early 1950s, pesticide use in the United States has increased tenfold, yet we still lose about a third of our crops to pests—the same proportion we lost before the pervasive use of chemicals. The actual percentage of vegetables lost to insects has doubled, mainly because the chemicals are indiscriminate in their killing. Birds, natural enemies of insects, also die, as do useful insects that eat other insects. The lower forms of life, including insects, have a unique ability to mutate and evolve within a few generations, producing offspring that are totally resistant to our poisons. Such was the case with DDT; the hardy *Anopheles* mosquito survived the DDT onslaught, and malaria, once thought to have been eliminated, is on the rise again.

Faced with insect resilience, the chemical companies keep providing new and even deadlier compounds to farmers and consumers, thus creating a toxic spiral in which the insects are the ultimate winners. Pesticides can cause cancer, birth defects, mutations, and sterility. We have only seriously tested 10% of them, but whether we test them or not, these poisons can hardly be expected to improve human health, and their cumulative and combined effects cannot be predicted.

Today many of the most dangerous pesticides, including DDT, aldrin, BHC, chlordane, and lindane, are banned in the United States, though their manufacture and export is not banned. Ironically, even DDT, which has been banned here for 19 years, is still used in Third World countries on coffee beans and tropical fruits that are then imported by the United States. Mexican fruits and vegetables, which make up 90% of the U.S. produce supply from December to May, contain 6 times the level of pesticide residue considered acceptable by the U.S. Department of Agriculture, which is helpless to control their import.

The pesticide DBCP (dibromochloropropane), used in the United States for the past 20 years, is now known to contaminate at least 750 wells in California's Central Valley, one of the largest agricultural areas in the country. Discovery of its presence in these wells shattered the belief that DBCP would evaporate or degrade and not migrate through the soil to the groundwater. A positive link has been established between DBCP contamination of drinking water and increased deaths from leukemia, stomach cancer, and thyroid cancer. In analyzing water from 3,000 California wells, the California Assembly Office of Research has also found 56 other pesticides.

A rodenticide known as EDB (ethylene dibromide) was used for 30 years as a fumigant in grain storage. A proven carcinogen, it was recently banned in the United States by the EPA. Yet trace amounts are still permitted, and this toxin shows up in our breakfast cereals, cake mixes, and bread.

Mercury

Mercury has entered the food chain in various parts of the world, notably the St. Clair River area of Michigan near the Canadian border and Minimata Bay in Japan. In these areas, industries discarded mercury wastes into the water, the mercury was absorbed by fish, and people who ate the fish developed severe neurologic and other disorders. Mercury is also found frequently in ocean waters and in the tissue of ocean fish like swordfish, shark, and tuna; the older and larger the fish, the higher the concentration of mercury. The mercury scare of the early 1970s involving tuna and swordfish has been forgotten, but the mercury remains. In the nineteenth century, felt hat manufacturers discarded mercury into rivers, polluting some water supplies; they themselves often suffered from mercury poisoning, which causes blind staggers—hence the expression "mad as a hatter."

Fire Retardants and Insulating Materials

Fire retardants known as PBBs (polybromated biphenyls) were accidentally shipped to farmers in Michigan in 1973 in place of animal feed. Masses of livestock were poisoned, and the farmers had to destroy over 25,000 dairy cows and huge numbers of poultry and eggs. The only way to get rid of the carcasses was to bury the animals in mass graves. However, PBBs are very stable, so they are still in the soil and will reenter the food chain at some point. Farmers and their families who ate some of the tainted meat experienced a variety of complaints, including chloracne, joint pains, neurologic difficulties, and an increased incidence of abortions and birth defects.

Far more common than the PBBs are the PCBs (polychlorinated biphenyls). This group of chemicals, widely used as fire retardants and insulating materials, has become one of the most feared of the toxic wastes. PCBs were first synthesized in 1881 and first utilized industrially 50 years later when their useful properties were recognized. Since 1929, 1.4 billion pounds of PCBs have been produced in the United States alone. Manufacture of these compounds was banned in 1977, but reliable estimates suggest that there may be 750 million pounds still in use. Although production of PCBs has ceased, the EPA has allowed their continued use in closed (theoretically leakproof) systems. Despite this restriction, significant amounts of PCBs have recently been found contaminating fuel oil shipped via pipeline in direct defiance of the EPA regulation.

The first major PCB accident occurred in Japan in 1968 when PCBs leaking from a heat exchanger contaminated rice oil. Over 1,200 people who consumed the oil became ill. They experienced such symptoms as swollen, painful joints, severe skin rashes, unusual discharges from the eyes, and gum discoloration. Since the early 1970s, there has been a PCB-related ban on sport and commercial fishing in the Hudson River in New York State. Because PCBs are so stable, tons of them will remain in the sediment of the Hudson and other American waterways for an untold period of time. PCBs have long since worked their way into the food chain, and the EPA has estimated that over 90% of all Americans have detectable traces of PCBs in their bodies.

Household Toxins

Various household products may cause a range of problems, some more serious than others. Penta (pentachlorophenol), used in homes and on playground equipment, decks, and picnic tables to preserve wood and prevent termite infestation, was linked to birth defects more than 12 years ago. Penta contains a form of dioxin that may cause cancer, and the EPA attempted in 1984 to prevent over-the-counter sale of the chemical. However, the decision is being appealed and is likely to be tied up for many years. Meanwhile, penta is still on the market. The only warning on its label states that it is toxic to fish.

Another dangerous chemical found in home-use products is captan, a close relative of the drug thalidomide, which caused a wave of birth defects some years ago. The safety tests that allowed captan to be placed on the market were proved to be fraudulent, those who perpetrated the fraud were sent to jail, but the deadly chemical is still available. It is used as a spray for grapes and garden tomatoes and as an ingredient in cosmetics, shampoo, and some foods. In 1980 the EPA began a study of captan that is expected to end as this book goes to press. Even if the EPA decides to ban captan, the appeals process will permit it to remain on the market for many years.

Another serious xenobiotic often found in the home is benomyl (butylcarbamoy-benzimidazole carbamate), which is used to kill fungus or mold on lawns and on peach crops prior to harvesting. Benomyl is linked to birth and genetic defects as well as infertility. Lindane, mentioned above among the pesticides, is used in many home products, including medicines, floor wax, and pet dip.

Nitrates

Nitrate contamination of groundwater is one of the fastest-growing toxin problems today. Such contamination is primarily due to runoff from agricultural fertilizer, chemical or natural, abetted by septic tank percolation and land disposal of organic wastes. The nitrate problem is another toxic time-bomb.

In the past, nitrate contamination was primarily a rural problem. While it is still true that the most heavily affected areas are agricultural, nitrate contamination has spread everywhere because of the growth of agribusiness. It has become the number one reason for shutting down public water wells in the United States.

The problem arises because nitrogen compounds used as fertilizer (ammonia, chemical preparations, animal waste producuts) are converted to nitrites and then to nitrates by aerobic bacteria in the soil. The nitrates and nitrites percolate down through the soil into the groundwater, often accompanied by bacteria. This affects the farm's well water and also the aquifers that supply cities near farming areas.

The mixture of nitrates, nitrites, and bacteria in water is particularly dangerous to infants under the age of six months, because their poorly developed gastrointestinal tracts allow the further conversion of nitrates into nitrites, which are able to combine vigorously with infant hemoglobin. The result is methemoglobinemia (one form of the phenomenon popularly known as "blue baby"), a condition that prevents the infant from utilizing oxygen readily and gives a bluish cast to the skin (cyanosis).

Unfortunately, boiling does not get rid of nitrates in water; in fact, it concentrates them. And it is just as important for nursing mothers to avoid nitrates as for babies, for nitrates can be passed on to the infant through the mother's milk.

The main problem with high nitrates is that they can be converted into nitrosamines in the gastrointestinal tract. Nitrosamines are xenobiotics that have been implicated in the development of birth defects, stomach cancers, and esophageal cancers.

THE X FACTOR IN THE FOOD CHAIN

The chemicals we are putting into the environment can show up in virtually anything we eat or drink—water, meat, fish, poultry, eggs, coffee, dairy products, alcoholic beverages, fruit, vegetables, grains. Some of our food sources, however, are more likely than others to carry large doses of toxins; as pointed out earlier, animals—from mollusks to humans—are bioconcentrators and biomagnifiers of the X factor.

Animals and humans absorb nutrients through the gastrointestinal tract, which is also the entry point for most of the modern-day xenobiotics. However, considerable amounts of such chemicals as solvents, chlorine, and chloramine can be absorbed through the skin, especially during industrial contact or prolonged bathing, swimming, or use of chlorinated spas. We also absorb various poisons by breathing.

Our gastrointestinal tract is designed to absorb or digest just about everything; for this reason it cannot serve as a defense against xenobiotics (nor can the gastrointestinal systems of the animals we eat). From the gastrointestinal tract food is absorbed into the bloodstream and goes directly to the liver, where a complex series of enzymes continues to process the fats, carbohydrates, and amino acids before they are distributed to the rest of the body for further metabolism. The liver, with its cytochrome P-450 complex of detoxifying enzymes, is also the first organ to encounter any xenobiotics ingested along with food.

The liver's numerous enzyme systems can often neutralize poisons in small amounts and arrange for them to pass out with bile juices or pass to the kidneys and out with the urine. But sometimes the liver is perplexed when it comes into contact with poisons it cannot defuse, and these end in the fatty tissue or cholesterol while the liver tries to come up with a strategy for handling them. The insecticide DDT is one such poison. Occasionally the liver is also unable to entirely detoxify certain xenobiotic drugs, which can cause Parkinson's disease, cirrhosis, systemic lupus, and other man-made diseases.

Sometimes the detoxification process backfires. The liver has just so many enzymes to deal with an increasing number of xenobiotics. Although insects are equipped to mutate rapidly and adapt to new poisons in a few generations, the same is not true for humans. Allow a few hundred thousand years, and our livers might evolve to cope with the poisons of progress over the past 200 years. But as it is, through a process called biotransformation, our uneducated liver enzymes may foul up the detoxification process and convert a relatively innocuous chemical into a toxic substance, usually a free radical. For example, the liver might take a stable chlorinated hydrocarbon compound and unknowingly convert it into a free radical that will attack DNA or other vital cell structures. This has been shown to occur with carbon tetrachloride and dibenzanthracene.

At present, scientists suspect that the highly reactive singlet oxygen and peroxide free radicals formed by oxygen metabolism contribute to the aging process and the development of several of the "degenerative" diseases, but it is important to broaden our thinking and investigate all the xenobiotics that enter the human body. While it is true that oxygen metabolism, a normal body process, generates free radicals as intermediate by-products, the countless toxins we ingest are much more likely suspects as causes of disease.

Most Americans consume in excess of five pounds of food additives each year. Thus, it is also important to question official reassurances that food additives "pose no threat" to us, that tiny amounts of captan will not be deleterious over a lifetime—or a little bit of FD&C #3 food dye each day, or "below-tolerance" levels of dieldrin or any of the other poisons that fill our stores and our homes.

MONITORING THE X FACTOR

The governments of the industrialized nations attempt in various ways to control the pollutants and poisons in food, water, and air, but most of these controls are ineffective. However, it would be unfair to accuse government officials of negligence, for most do make a valiant effort within the constraints imposed upon them by commercial interests, bureaucracy, and inadequate laws. In the United States, there are two important agencies concerned with the regulation of chemical pollutants: the Environmental Protection Agency and the Food and Drug Administration.

The EPA was formed as an independent part of the executive branch in 1970 by President Richard Nixon. Its duties include air pollution control, solid waste management, drinking water quality control, water pollution oversight, and the establishment of environmental radiation protection standards and tolerance levels for pesticides.

Part of the EPA's mandate is setting standards of water quality. However, the mere existence of standards does not necessarily result in significant improvements. Pollutants in the waterways have not been eliminated, just regulated. Industries and municipalities have not stopped discharging waste; they can continue to do it as long as they have a permit. The permit is supposed to insure that the waste is treated before disposal, but this is not easily monitored, and even treated waste is suspect. There are also polluters who manage to escape the

EPA's notice. In addition, the amounts of pollution discharged legally are often much higher than an already polluted body of water can tolerate. This is especially true in industrialized urban areas. The EPA is virtually powerless to deal with previously contaminated rivers, streams, and aquifers.

Some headway has been made against air pollution in the last ten years, including a decrease in levels of ambient lead, ozone, carbon monoxide, and sulfur dioxide. However, even with this modest improvement, acid rain from industrial emissions continues to be a major problem in the United States, Canada, and Europe.

Americans are hooked on disposables. For years, we have discarded boxes, containers, old furniture, paint cans, aerosol cans, and countless other items no longer of use—billions of tons of waste. In most cases, this waste has been buried in the ground in "safe" sanitary landfills—which have proven unsafe after all, as they frequently leach into underground water aquifers. The problem of solid waste disposal is certain to grow worse with time; toxic waste landfills have already produced several catastrophes, notably Love Canal in New York and the Stringfellow Acid Pits near San Bernardino, California. We need new technology to dispose of wastes, not new burial sites that will contaminate fragile soil and water aquifers.

The U.S. Congress has established a Superfund to be used by the EPA for the cleanup of contaminated areas. However, the EPA bureaucracy generally delegates investigation and cleanup to state officials and local EPA offices. The amount of money spent and the effectiveness of cleanup operations, then, depends upon the vigor and ambition of these local groups. At this point, 850 sites across the nation have been identified as dangerous enough to qualify for Superfund money. However, according to the EPA, there are 19,000 sites that are potentially hazardous enough to qualify, and the congressional Office of Technology Assessment says that there may actually be as many as 100,000.

The passage of The Pure Food and Drug Act in 1906 permitted the federal government to regulate food and drugs for the first time. It was an important step in rectifying the unsanitary conditions commonplace in the burgeoning food processing industry, and in cracking down on worthless nostrums, patent medicines, and dangerous drugs such as opium, cocaine, heroin, and morphine. The act came about primarily through the efforts of Dr. Harvey W. Wiley, chief of the Bureau of Chemistry of the Department of Agriculture. His department administered the law until 1927, when the Food, Drug, and Insecticide Administration was formed. It was ultimately renamed the Food and Drug Administration and was transferred from the Department of Agriculture to what is today the Department of Health and Human Services.

For its time, the Pure Food and Drug Act was a strong law, but with the rapid emergence of technology in the twentieth century, it quickly became outdated. Amendments to the law became necessary to increase the authority of the FDA. After diethylene glycol, a poisonous solvent in sulfanilamides, caused the deaths of 107 people, mostly children, the Federal Food, Drug, and Cosmetic Act was passed in 1938 to regulate drugs prior to marketing. This act also broadened the FDA's mandate: cosmetics and medical devices were now to be regulated; false claims for drugs could be suppressed without proof of intent to defraud; drug manufacturers were required to provide scientific proof that new products were safe before putting them on the market; and the addition of poisonous substances to foods was prohibited or, where unavoidable or required in production, restricted to prescribed "tolerance" levels.

The thalidomide tragedy of the early 1960s brought about additional changes in drug regulation. Thalidomide was a mild sedative declared safe after stringent, long-term testing on 35,000 animals. It was sold without prescription in Europe and was made available with FDA approval to about 1,200 American physicians for clinical trials. Thalidomide remains one of the most vivid examples of the inadequacy of testing as a method of detecting drug toxicity.

Most people in the United States assume that they are protected by the FDA and the EPA. However, that belief should be examined. The EPA and the FDA see no harm in the fact that the standards they set for bottled water companies are no better than tap water standards; neither agency considers chlorine and the trace elements absorbed by water from pipes and surrounding soil to be a major problem.

A reading of the FDA laws on food additives also gives one pause. According to the FDA definition, "Food additives are substances which by their intended use may become components of food, or which otherwise affect the characteristics of the food." Specifically exempt by law from the definition are the following: "(1) substances generally recognized as safe by qualified experts [the GRAS list includes such "safe" chemicals as the sulfiting agents, which have since been implicated in numerous allergic reactions and several deaths] (2) substances used in accordance with a previous approval; (3) pesticide chemicals in or on raw agricultural products; (4) a color additive; and (5) a new animal drug."

The FDA does not usually undertake the testing of new drug preparations but only evaluates the results in deciding whether to approve the drug. Researchers from universities or pharmaceutical companies generally do the testing, first on animals and eventually on humans. Although the FDA is concerned about drug safety, experience has shown that even drugs believed to be safe can have disastrous consequences. There are also numerous instances of unscrupulous researchers and companies falsifying data.

With so much potential for outright fraud or disaster through error, it is no wonder that the EPA and the FDA have not given much consideration to subtle low-grade poisons that may have long-range effects. Rather than require the elimination of many of these poisons from food and water, both the EPA and the FDA have set up tolerance levels. In many instances the determination of "safe" levels of poisons is not based on reliable scientific evidence but on guesswork or estimates from incomplete animal studies.

When it is finally recognized that current tolerance levels for pesticides and other poisons may be too high to prevent long-term damage, perhaps we can expect a change in attitude. The FDA considers that it "has a double responsibility to protect the public from harm and to encourage technological advances that hold promise of benefits to society." We must bear in mind, however, that technological innovations have been responsible for both good and bad, and will probably continue to be so. We must reassess the damage that inevitably results from the dispersal of chemicals into our fragile environment, and we must find new ways to contain, reuse, or eliminate them. Above all, we must insure that technological advances are truly advances, demonstrably safe now and in future generations, and that their "promise of benefits to society" is a promise they can keep.

Significant Developments in the Industrial Revolution

Date	Development
1700	Rolling mill
1701	Seed drill
1709	Coke-fired blast furnace
1714	Horse hoe
1733	Flying shuttle; blood pressure of horse recorded
1746	Manufacture of sulfuric acid by lead-chamber process
1751	Crucible process for casting steel
1759	Practical use of coke-fired blast furnace; marine chronometer
1760	Drill and lathe
1764	Spinning jenny
1765	Solvent for rubber
1769	Waterframe spinning machine
1770	Grand Trunk Canal, England
1772	Double-acting expansive engine
1774	Boring mill
1776	Steam engine
1778	Water closet

Date	Development
1781	Epicyclic gear
1783	First balloon ascent
1784	Spinning mule; reverberatory furnace for wrought iron
1785	Chemical bleaching with chlorine; power loom
1789	Leblanc process for obtaining soda from salt
1792	First manufacture of window glass in U.S.
1793	Cotton gin
1796	Hydraulic press; smallpox vaccine
1797	Screw-cutting lathe
1800	Interchangeable parts; galvanic cell (electric battery)
1801	Sugar beet cultivation; Jacquard loom; aqueduct water in Philadelphia
1802	High-pressure steam engine; first sheet copper produced in U.S.; steam carriage
1804	First locomotive on rails
1806	Gas lighting of cotton mills
1807	Steamboat
1808	Anthracite coal burned in open grate
1810	Power-driven press; canning using glass jars
1811	Flatbed press; canning using tin cans
1812	Lead pencil manufacture
1813	Power loom
1814	First locomotive; first steam-powered warships; first factory for manufacture of cotton cloth by power machinery
1817	Machine-made paper
1818	Manufacture of patent leather in U.S.
1819	Stethoscope; cast-iron plow; first canning business in U.S. (fish)
1821	First natural gas production in U.S.
1822	False teeth
1823	Hydrofluoric acid
1824	First laboratory for teaching chemistry; Portland cement; the mackintosh
1825	First successful railroad system; mechanical pressing of glass
1826	First patent for internal-combustion engine
1827	Water turbine; manufacture of sulfuric acid by Gay-Lussac tower
1828	Synthesis of urea; American production of china
1831	Electric bell; power knitting machine
1832	Mechanical generation of electricity

Date	Development
1834	Vapor compression engine; McCormick reaper patented
1835	Colt revolver
1836	Galvanized iron
1837	Telegraph; steel plow; shell gun; differentiation of typhus from typhoid fever
1839	Vulcanized rubber; steam hammer; drop forging; dye stamping; daguerreotype photographic process
1841	Paper positives for photos; commercial use of oil as patent medicine in U.S.
1842	Superphosphates; first commercial artificial fertilizer from rock phosphate and sulfuric acid; ether as anesthesia
1843	First soap powder
1844	Gas engine using turpentine for fuel
1845	Brussels power loom for carpets; breech-loading artillery; air refrigerator
1846	Nitroglycerine; rotary press; lock-stitch sewing machine
1847	First adhesive postage stamp; revolving disk harrow
1848	First air conditioning installed in theater
1849	Minié bullet; safety pin
1850	Cable plow
1851	Wet collodion plate process; first electric fire alarm system in U.S.; first patent for ice-making machine
1852	Hydraulic elevators; first effective fire engine
1854	Kerosene
1855	Bunsen burner; turret lathe; first commercial oil business
1856	Bessemer process for converting pig iron into steel; first coal-tar dye (mauve); borax found in U.S.; first practical folding machine
1857	Fermentation scientifically explained
1858	Steam plow
1859	First oil well drilled; first passenger elevator in a hotel; first demonstration of electric home lighting
1860	Combine harvester; evaporated milk; ammonia absorption machine; first repeating rifle produced in U.S.
1861	Pasteurization of wine, beer, milk; ammonia process for manufacturing soda
1862	Universal milling machine; nitrated cellulose; Gatling machine gun
1863	Paper dress patterns sold; open-hearth process for steel

Date	Development
1864	Self-propelled torpedo; first salmon cannery in U.S.
1865	Antiseptic surgery; web press; compression ice machine; high-vacuum mercury pump; first U.S. company to sell natural gas
1867	Dynamite; velocipede; cigarettes introduced to U.S.
1868	Pork packing in U.S.; air brake; patent for refrigerator car
1869	Vacuum cleaner; beef packing in Chicago
1870	Ring armature; first practical dynamo; patent for celluloid
1872	Commercial production of celluloid
1873	Use of electricity to drive machinery; ammonia compression refrigerator; typewriter; barbed wire
1874	DDT synthesized; streetcar invented
1875	Blasting gelatin (gelignite)
1876	First practical gas engine; carpet sweeper; ammonia compressor engine; telephone; mimeograph
1877	Resistance welder; cream separator
1878	Sheaf-binding harvester; carbon filament electric lamp; first electric light company in U.S.; phonograph
1879	Incandescent bulb
1880	Chemical fertilizers; roll film; first municipal electric light plant; first standardized manufacture of house paint
1881	First central electric light power plant
1882	Electric generating stations; flat iron; electric fan
1883	Fountain pen; first cigar-rolling machine
1884	Steam turbine; smokeless gunpowder
1885	Wilshaw gas mantle; chain drive bicycle; Daimler gasoline engine; furnaces for garbage disposal; Benz automobile using internal-combustion engine
1886	Hydroelectric installation at Niagara Falls; electrolytic method of obtaining aluminum from bauxite
1888	Alternating current; electric motor; adding machine; hand camera; pneumatic tire
1889	Cordite; spindle-type cotton picker; Bessemer steel; I-beams
1890	Concept of immunization; cyanide process for extracting gold; flue curing of tobacco leaf; saccharin
1891	Sulfur mined by superheated steam; zipper patented
1892	Diesel heavy oil engine; gasoline tractor; first comprehensive work in bacteriology; first gasoline and first electric auto; book matches
1893	Velox paper for photography

Date	Development
1895	Motion pictures; wireless telegraph; safety razor; liquid air
1896	Electric stove; first practical sphygmomanometer; X-ray treatment for breast cancer
1897	Cathode-ray tube
1898	Magnetic recording of sound; bottle-making machine; quick-firing artillery piece
1900	Oxyacetylene torch; Zeppelin airship
1901	Mercury vapor lamp
1902	Rayon (first cellulose fiber); hydrogenation of fat for soap and foods
1903	Airplane
1904	Radio tube; yellow fever study in Panama
1906	Neoprene (synthetic rubber); electrode vacuum tube (electronics)
1907	Washing machine
1908	Lethal gas recommended for military use
1909	Bakelite (first polymer thermosetting plastic); chemotherapy treatment for syphilis
1911	Electric self-starter for automobiles
1912	Cellophane; discovery of vitamin A and thiamine;
1913	Mass production of the zipper; diesel electric railway engine
1914	Packaged prerolled cigarettes; conveyor-belt mass production; successful heart surgery on an animal
1916	Submachine gun
1917	Freezing of foods
1919	First municipal airport in U.S.
1920	Arc welding; commercial radio broadcasting
1921	Tetraethyl lead gasoline additive
1922	Technicolor for motion pictures; mechanical switchboards; first fruitful radar research in U.S.; commercial production of cellophane; discovery of leukocytes
1925	Dry ice; scarlet fever antitoxin
1926	Sound motion pictures; television; first successful treatment of pernicious anemia
1927	Iron lung
1928	Electric razor; first animated electric sign
1929	Frozen foods sold commercially; coin-operated vending machines
1930	Gas turbine for jet aircraft

Date	Development
1931	Catalytic cracking system for petroleum
1932	Polaroid glass
1933	Fluorescent lamp; frequency modulation (FM)
1934	Nylon
1935	Plastic tape recording; color film; sulfa drugs
1937	Xerography; National Cancer Institute founded
1938	Nylon patented; fiberglass; ballpoint pen; radar system demonstrated; fluoride recommended for water supply
1939	DDT used as an insecticide; nylon stockings sold; U.S. turbojet airplane; helicopter; discovery of Rh factor in blood; penicillin
1940	Artificial insemination of livestock; electron microscope
1941	Acrilan, Orlon, Dynel, Dacron
1942	Rockets and missiles; first American jet airplane; extensive development of oil pipelines; first nuclear chain reaction
1943	Large-scale production of penicillin
1944	Digital computer; synthetic quinine; streptomycin
1945	Silicon polymers; industrial development of silicone; fluoridation of water supplies; plastic packaging of food; discovery of folic acid
1946	Hydrocarbon dating
1947	Polaroid camera; supersonic airplane; dry-ice seeding of clouds
1948	LP phono recordings; transistors; Aureomycin; vitamin B-12
1949	Cortisone; commercial use of detergents
1950	Mass production of poultry; nuclear propulsion of submarines; color television broadcast
1951	First electricity produced by atomic energy
1952	Hydrogen bomb
1953	Electronic computer model for DNA; polio vaccine
1954	Solar battery; maser
1955	Thorazine and reserpine used on mental patients
1956	Nuclear fuel used for electricity
1957	First satellite in space
1958	Jet-powered transatlantic airline service; link between carbon 14, cancer, and leukemia established
1960	Laser

Directory of Selected Resources

In order to remain healthy, *you* must assume the major responsibility for your life. Following the measures outlined in this book is a crucial step in that direction. It is also part of your responsibility to do your own research and reading. However, there will still be times when you need the services of a caring and knowledgeable professional. If you are ever in doubt about the advice your doctor gives you, it is up to you to seek help elsewhere. Please don't be concerned about hurting your doctor's feelings; no caring doctor would place his pride above your welfare.

Listed below, alphabetically, are some publications and organizations dealing with patient care and education on a variety of topics. Their philosophies range over a wide spectrum, traditional and nontraditional. Some of the organizations may be able to refer you to a doctor interested in preventive medicine. These are all reputable groups, but you will want to check them out to see if they fit your needs.

American Academy of Medical Preventics, 6151 W. Century Blvd., Suite 1114, Los Angeles, CA 90045 (213-655-4310). Though many of the ideas of this organization appeal to me, their major thrust is not preventive medicine per se but the use of chelation therapy.

Many people believe in chelation. I have not found enough evidence to recommend it, nor can I unequivocally dismiss it. I would reserve judgment for the time being.

American Holistic Medical Association, 2002 Eastlake Ave. East, Seattle, WA 98102 (206-322-6842). Holistic medicine takes on various shades, depending on the practitioner. This organization maintains a list of physicians who practice preventive medicine. It will also supply you with various types of literature, including a newsletter.

Center for Science in the Public Interest, 1501 16th St. NW, Washington, D.C. 20036-1499. CSPI is the coordinator of a group called Americans for Safe Food, which is supported by many consumer and environmental organizations. Their goal is to organize a grass-roots movement that will encourage the widespread availability of contaminant-free foods, fight for disclosure of food additives, demand a ban on pesticides and animal drugs, and press for national standards on "organic" and "natural" foods. You can join CSPI and receive their excellent *Nutrition Action Health Letter* for an annual membership fee of $19.95.

Feingold Associations of the United States, PO Box 6550, Alexandria, VA 22306 (703-768-FAUS). FAUS sponsors local support groups throughout the United States for parents and relatives of children with behavior and learning disorders. Its purpose is to generate public awareness of the role played by synthetic colors, synthetic flavors, and the preservatives BHA, BHT, and TBHQ in the genesis of hyperactivity, learning disorders, and behavior disorders in children. FAUS believes that nutrition, rather than drugs, is the way to solve most behavioral problems. Membership entitles you to their newsletters, a handbook for following the Feingold program, and a food list by brand name with specific shopping advice. FAUS holds regularly scheduled meetings and sponsors a help hotline.

Human Ecology Action League, PO Box 1369, Evanston, IL 60204. HEAL publishes a quarterly magazine, *The Human Ecologist*, and sponsors local support groups throughout the country for people suffering from environmentally related disorders. Much of what this organization does and says makes an extremely important contribution, though I do find their overemphasis on *Candida albicans* a bit disconcerting.

Kushi Institute, PO Box 1100, Brookline, MA 02147 (617-731-0564). Mishio Kushi is the foremost authority on the macrobiotic diet. His institute can teach you how to prepare foods the macrobiotic way. If I am not mistaken, macrobiotics works for the same reasons the low-toxin program works.

The McDougall Newsletter, PO Box 1761, Kailua, HI 96734. This excellent bimonthly newsletter is written by John A. McDougall, M.D., author of two fine books, *The McDougall Plan* and *McDougall's Medicine*. Subscription cost is $8.00 per year.

Medical Center for Health and Longevity, 23560 Madison St., Suite 102, Torrance, CA 90505 (213-539-1501). This is my organization for research and education. It publishes a newsletter, *Ten-Point Bulletin*, containing updates on continuing research, water quality, antioxidants, preventive measures, etc. As a reader of this book, you are entitled to a free issue. You can also contact me at the Center for recipes or further information on the ideas in this book.

National Academy of Clinicians and Holistic Health, PO Box 271, New York, NY 11422. This organization promotes a holistic approach to health and the clinical use of "natural" medicine, acupressure, and related practices. It sponsors a speakers' bureau and offers books, tapes, and a directory of physicians and allied health professionals.

National Coalition Against the Misuse of Pesticides, 530 7th St. SE, Washington, D.C. 20003 (202-543-5450). NCAMP's goal is to publicize the widespread pesticide contamination of food and water supplies and to promote alternative pest control strategies that reduce or eliminate the use of chemicals. The organization publishes a monthly newsletter that will keep you informed about pesticide-related legislation and other developments.

National Women's Health Network, 224 7th St. SE, Washington, D.C. 20003. "The nation's only consumer organization devoted to women's health issues" maintains a large resource library and publishes a newsletter. Membership dues are $25.00 per year.

People's Medical Society, 33 E. Minor St., Emmaus, PA 18049. This organization was established to promote preventive medical practices and contain the cost of medical care. It is primarily a referral service. It does not maintain a list of recommended physicians, but it can refer you to groups that do. If you have a problem with a medical bill or a dispute with a doctor, hospital, or insurance company, it will advise you or direct you to an appropriate agency or professional.

Professional Books, 681 Skyline Dr., Jackson, TN 38301 (800-DR-CROOK). If you suspect that monilia or *Candida albicans* is your problem, there is no better source of information than my friend Dr. William Crook, author of *The Yeast Connection*. His organization will supply you with literature, tapes, and books dealing with this and related problems. Personally, I am not convinced that *Candida* is a major problem for most people; my own view is that xenobiotic damage to the immune system is far more common and is the reason why yeast and other opportunistic infections take hold.

Tecbook Publications, PO Box 5002, Topeka, KS 66605. A good source of inexpensive (sometimes free) pamphlets, newsletters, and booklets dealing with traditional and alternative medical care.

Vegetarian Times, 141 S. Oak Park Ave., PO Box 570, Oak Park, IL 60303 (312-848-8120). This magazine is a good source of new recipes, as well as current information on diet and health. Remember, however, that vegetarians are not all alike; some of the recipes may contain dairy products, eggs, and other ingredients that you should avoid on a low-toxin diet. Annual subscription cost is $19.95 for 12 issues.

Bibliography

This bibliography includes references for the material in each chapter, in the same order as the material is presented. Asterisks indicate sources that are of particular value for additional reading but were not specifically used for the text.

Introduction

Eaton, S. B., and Konner, M. "Paleolithic Nutrition: A Consideration of Its Nature and Current Implications." *New England Journal of Medicine* 312 (1985):283–89.

Osborne, T. "Amino Acids in Nutrition and Growth." *Journal of Biological Chemistry* 17 (1914):325–28.

Lappé, F. M. *Diet for a Small Planet.* New York: Ballantine Books, 1971.

*Snowden, D. A. "Epidemiology of Aging: Seventh-Day Adventists—A Bellwether for Future Progress." In *Intervention in the Aging Process*, edited by W. Regelson and F. M. Sinex, pp. 141–49. New York: Alan R. Liss, 1983.

Sacks, F. M., et al. "Plasma Lipids and Lipoproteins in Vegetarians and Controls." *New England Journal of Medicine* 292 (1975):1146–51.

Armstrong, B., et al. "Blood Pressure in Seventh-Day Adventist Vegetarians." *American Journal of Epidemiology* 105 (1977): 444–49.

McLaren, D. "The Great Protein Fiasco." *Lancet* 2 (1947):93–97.

Part I: A NEW APPROACH TO HEALTH AND LONGEVITY

Chapter 1: The X Factor

*St. George, D. "Life Expectancy, Truth, and the ABPI." *Lancet* 2 (1986):346.

McKeown, T. *The Role of Medicine: Dream, Mirage, or Nemesis?* London: Nuffield Provincial Hospitals Trust, 1976.

Lyons, A. S., and Petrocelli, R. J., II. *Medicine: An Illustrated History.* New York: Harry N. Abrams & Co., 1978.

Pryor, W. A., ed. *Free Radicals in Biology*, vol. 6. Orlando, Fla.: Academic Press, 1984.

Armstrong, S., et al. *Aging.* Vol. 27, *Free Radicals in Molecular Biology, Aging, and Disease.* New York: Raven Press, 1984.

Burnum, J. F. "Medical Vampires." *New England Journal of Medicine* 314 (1986): 1250–51.

*Stini, W. A. "Early Nutrition, Growth, Disease, and Human Longevity." *Nutrition and Cancer* 1 (1978):31–39.

*Ross, M. H., and Bras, G. "Lasting Influence of Early Caloric Restriction on Prevalence of Neoplasms in the Rat." *Journal of the National Cancer Institute* 47 (1971):1095–1104.

Mason, R. P. "Free Radical Metabolites of Toxic Chemicals." *Federation Proceedings* 45 (1986):2464.

*Walford, Roy L. *Maximum Life Span.* New York: W. W. Norton & Co., 1983.

Schneider, E. L., and Reed, J. D., Jr. "Life Extension." *New England Journal of Medicine* 312 (1985):1159–65.

*Porter, N. A. "Endoperoxides and Intermediates in Prostaglandin (PG) Biosynthesis: Chemical Theory Relating to the Molecular Basis of Heart Attack and Stroke." Chap. 8 in *Free Radicals in Biology*, vol. 4, edited by W. A. Pryor. Orlando, Fla.: Academic Press, 1980.

*Leaf, Alexander. "Getting Old." *Scientific American* 229 (1973):45–52.

Leaf, A., and Launois, J. "Search for the Oldest People: Every Day Is a Gift When You Are Over 100." *National Geographic* 143, no. 1 (1973):92–119.

*Comfort, A. *The Biology of Senescence.* New York: Elsevier, 1978.

*Weindruch, R. H., et al. "Modification of Age-Related Immune Decline in Mice Dietarily Restricted from or after Mid-Adulthood." *Proceedings of the National Academy of Science, U.S.A.* 79 (1982):898.

*Pitskhelauri, G. Z. *The Long-Living of Soviet Georgia.* New York, Human Sciences Press, 1982.

Chapter 2: Combating the X Factor

Trowell, H. C., and Burkitt, D. P. *Western Diseases: Their Emergence and Prevention.* London: Edward Arnold, 1981.

Burkitt, D. "Some Diseases Characteristic of Modern Western Civilization." *British Medical Journal* 1 (1973):274–81.

Burkitt, D. P., et al. "Effect of Dietary Fibre on Stools and Transit-Times and Its Role in the Causation of Disease." *Lancet* 2 (1972):1408–12.

Trowell, H. C. "Dietary Fibre and Colonic Disease." In *The Present State of Knowledge*, no. 6. London: Norgine Ltd., 1974.

Trowell, H. C. "Dietary Fibre: Metabolic and Vascular Diseases." In *The Present State of Knowledge*, no. 7. London: Norgine Ltd., 1975.

James, W. P. T., et al. "Calcium Binding by Dietary Fibre." *Lancet* 1 (1978):638–39.

Anderson, James W. "Physiologic and Metabolic Effects of Dietary Fiber." *Federation Proceedings* 44 (1985):2902–6.

Prins, R. A. "Biochemical Activities of Gut Microorganisms." In *Microbiology of the Gut*, edited by R. T. Clarke and T. Bauchop. New York: Academic Press, 1977.

Reddy, B. S., et al. "Promoting Effect of Sodium Deoxycholate on Colon Adenocarcinomas in Germ-Free Rats." *Journal of the National Cancer Institute* 56 (1976):441–42.

Lehmingen, A. L. *Biochemistry.* New York: Worth Publishers, 1970.

Hegenrather, J., et al. "Pollutants in Breast Milk of Vegetarians." *New England Journal of Medicine* 304 (1981):792–96.

U.S. Congress. Senate. *Diet Related to Killer Diseases.* 4 vols. Hearings Before the Select Committee on Nutrition and Human Needs, March 24, 1977. 95th Cong., 1st sess.

Hesse, F. G. "Incidence of Disease in the Navajo Indian." In *Health Problems of U.S. and North American Indian Populations*, pp. 157–61. New York: MSS Information Corp., 1972.

Kositchek, R. J., et al. "Biochemical Studies in Full-Blooded Navajo Indians." *Circulation* 23, no. 2 (1961):219–24.

Kunitz, S. J. *Disease Change and the Navajo Experience*. Berkeley: University of California Press, 1983.

Darby, W. J., et al. "Study of Dietary Background and Nutrition of Navajo Indians." *Journal of Nutrition* 60, supp. 2 (1956):1–85.

Davidson, Bill. "What Can We Learn About Health from the Mormons?" *Family Circle*, January 1976.

Pritikin, N., and McGrady, P. M., Jr. *The Pritikin Program for Diet and Exercise*. New York: Grosset & Dunlop, 1979.

Sattilaro, A., and Monte, T. *Recalled by Life*. Boston: Houghton Mifflin, 1982.

Dock, William. "The Reluctance of Physicians to Admit That Chronic Disease May Be Due to Faulty Diet." *Journal of Clinical Nutrition* (March 1953).

Dickinson, F. G., and Martin, L. W. "Physician Mortality, 1949–1951." *Journal of the American Medical Association* 162 (1956):1462–68.

Rimpelä, A. H., et al. "Mortality of Doctors: Do Doctors Benefit from Their Medical Knowledge?" *Lancet* 1 (1987):84–87.

Chapter 3: Preparing to Begin the Program

"Survey of Physicians' Attitudes and Practices in Early Cancer Detection." *Ca—A Cancer Journal for Clinicians* 35, no. 4 (1985):197–213.

Wald, N. J. "Mortality from Lung Cancer and Coronary Heart Disease in Relation to Change in Smoking Habits." *Lancet* 1 (1976):136–38.

Wynder, E. L., and Lemon, F. R. "Cancer, Coronary Artery Disease and Smoking: A Preliminary Report on Differences in Incidence Between Seventh-Day Adventists and Others." *California Medicine* 89 (1958):267–72.

Fielding, J. E. "Smoking: Health Effects and Control." *New England Journal of Medicine* 313 (1985):491–98.

Lemon, F. R., and Kuzma, J. W. "A Biologic Cost of Smoking: Decreased Life Expectancy." *Archives of Environmental Health* 18 (1969):950–55.

Aronow, W. S. "Effect of Passive Smoking on Angina Pectoris." *New England Journal of Medicine* 299 (1978):21–24.

Cahan, W. G. "Sins of Smoking Parents Against Their Embryos, Infants, and Children." *Medical Tribune*, April 3, 1985.

Greenberg, R. A., et al. "Measuring the Exposure of Infants to

Tobacco Smoke: Nicotine and Cotinine in Urine and Saliva." *New England Journal of Medicine* 310 (1984):1075–78.

Hirayama, T. "Non-Smoking Wives of Heavy Smokers Have a Higher Risk of Lung Cancer: A Study from Japan." *British Medical Journal* 282 (1981):183–85.

Matsukura, S., et al. "Effects of Environmental Tobacco Smoke on Urinary Cotinine Excretion in Nonsmokers." *New England Journal of Medicine* 311 (1984):828–32.

Weiss, S. T., et al. "The Health Effects of Involuntary Smoking." *American Review of Respiratory Diseases* 128 (1983):933–42.

Weiss, S. T., et al. "Persistent Wheeze: Its Relation to Respiratory Illness, Cigarette Smoking, and Level of Pulmonary Function in a Population Sample of Children." *American Review of Respiratory Diseases* 122 (1980):697–707.

White, J. R., and Froeb, H. F. "Small-Airways Dysfunction in Non-smokers Chronically Exposed to Tobacco Smoke." *New England Journal of Medicine* 302 (1980):720–23.

U.S. Department of Commerce. Bureau of the Census. *Statistical Abstract of the United States.* Washington, D.C.: Government Printing Office, 1985.

Knochel, James P. "Cardiovascular Effects of Alcohol." *Annals of Internal Medicine* 98 (1983):849–54.

Klatsky, Arthur L. "The Relations of Alcohol and the Cardiovascular System." *Annual Reviews of Nutrition* 2 (1982):51–71.

Popham, R. E., et al. "Heavy Alcohol Consumption and Physical Health Problems: A Review of the Epidemiologic Evidence." In *Research Advance in Alcohol and Drug Problems*, edited by R. G. Smart et al., pp. 149–82. New York: Plenum, 1984.

Part II: THE PROGRAM

Chapter 4: Water and Beverages

Sadler, T. D., et al. "The Use of Asbestos-Cement Pipe for Public Water Supply and the Incidence of Cancer in Selected Communities in Utah." *Journal of Community Health* 9, no. 4 (1984):285–93.

Harris, R. H. "The Implications of Cancer-Causing Substances in Mississippi River Water." Washington, D.C.: Environmental Defense Fund, 1974.

Page, T., et al. "Drinking Water and Cancer Mortality." *Science* 193 (1976):55–57.

Ward, P. S. "Carcinogens Complicate Chlorine Question." *Journal of the Water Pollution Control Federation* 46 (December 1974):2638–40.

Olivieri, V. P., et al. "Reaction of Chlorine and Chloramines with Nucleic Acids Under Disinfection Conditions." Chap. 57 in *Water Chlorination: Environmental Impact and Health Effects*, vol. 3. Ann Arbor, Mich.: Ann Arbor Science Publishers, 1980.

Francko, D. A., and Wetzel, R. G. *To Quench Our Thirst*. Ann Arbor, Mich.: University of Michigan Press, 1983.

Dorsch, M. M., et al. "Congenital Malformations and Maternal Drinking Water Supply in Rural South Australia: A Case-Control Study." *American Journal of Epidemiology* 4 (1984):473–86.

Mason, T. J., et al. "Asbestos-like Fibers in Duluth Water Supplies." *Journal of the American Medical Association* 228 (1974):1019–20.

Williamson, S. A. "Epidemiologic Studies on Cancer and Organic Compounds in U.S. Drinking Water." *Science of the Total Environment* 18 (1981):187–203.

Miller, R. W. "Carcinogens in Drinking Water." *Pediatrics* 57 (1976):462–64.

Maugh, T. N. "New Study Links Chlorination and Cancer." *Science* 211 (1981):694.

Carlo, G. L., and Mettlin, C. J. "Cancer Incidence and Trihalomethane Concentration in a Public Drinking Water System." *American Journal of Public Health* 70 (1980):523–24.

Robertson, J. S. "Minerals and Mortality." *Journal of the American Water Works Association* 71 (1979):408–13.

Waldbott, G. L., et al. *Fluoridation: The Great Dilemma*. Lawrence, Kans.: Coronado Press, 1978.

Richmond, V. L. "Thirty Years of Fluoridation: A Review." *American Journal of Clinical Nutrition* 41 (1985):129–38.

Bergland, F., et al. "Environmental Health Criteria, #36: Fluorine and Fluorides." Geneva: World Health Organization, 1984.

Lawrence, C. E., et al. "Trihalomethanes in Drinking Water and Human Colorectal Cancer." *Journal of the National Cancer Institute* 72 (1984):563–68.

Comstock, G. W., et al. "Water Hardness at Home and Death from Arteriosclerotic Heart Disease in Washington County, Maryland." *American Journal of Epidemiology* 112 (1980):209–16.

Punsar, S. "Cardiovascular Mortality and Quality of Drinking Water." *Work-Environment-Health* 10 (1973):107–25.

Sharrett, A. R., and Feinleib, M. "Water Constituents and Trace

Elements in Relation to Cardiovascular Disease." *Preventive Medicine* 4 (1975): 20–36.

*Helms, C. M., et al. "Legionnaires' Disease Associated with a Hospital Water System: A Cluster of 24 Nosocomial Cases." *Annals of Internal Medicine* 99 (1983):172–78.

Muss, D. L. "Relationship Between Water Quality and Deaths from Cardiovascular Diseases." *Journal of the American Water Works Association* 54 (1962):1371–78.

*"Water Supply and Environmental Sanitation." In *Guide to Health and Welfare Services in Japan*. Tokyo: Ministry of Health and Welfare, 1979.

Gottlieb, M. S ., et al. "Drinking Water and Cancer in Louisiana: A Retrospective Mortality Study." *American Journal of Epidemiology* 116 (1982):652–57.

Gaston, J. F. "Contamination of Drinking Water." *EPA Journal* 10, no. 6 (1984):20–21.

Environmental Protection Agency. Water Programs. "National Interim Primary Drinking Water Regulations." *Federal Register*, Part 4, March 12, 1982.

Studlick, J. R., and Bain, R. C. "Bottled Water: Expensive Ground Water." *Water Well Journal* 34, no. 7 (1980):15–79.

Baker, M. N. *The Quest for Pure Water*, vol. 1. Denver, Colo.: American Water Works Association, 1948.

Elwood, P. C., et al. "Mortality in Adults and Trace Elements in Water." *Lancet* 2 (1974):1470–72.

Clifford, D. "Point-of-Use Treatment for Nitrate Removal." *Water Technology* 7, no. 8 (1984):28–36.

*Luoma, J. R. *Troubled Skies, Troubled Waters: The Story of Acid Rain*. New York: Viking Press, 1984.

*McDaniels, A. *Water: What's in It for You*. San Pedro, Calif.: Heather Foundation, 1972.

Carmichael, N. G., et al. "Minireview: The Health Implications of Water Treatment with Ozone." *Life Science* 2 (1982):117–29.

Maruoka, S., and Yamanaka, S. "Mutagenic Potential of Laboratory Chlorinated River Water." *Science of the Total Environment* 29 (1983):143–54.

Beresford, S. A. "Cancer Incidence and Reuse of Drinking Water." *American Journal of Epidemiology* 117 (1983):258–68.

Environmental Protection Agency. Criteria and Standards Division, Office of Drinking Water. "Bottled Water." Bulletin WH-550.

Environmental Protection Agency. "Preliminary Assessment of Carcinogens in Drinking Water." Report to Congress, 1975a.

California. Department of Health, Water Sanitation Section. "Wastewater Reclamation Criteria." From California Administrative Code Title 22, Division 4, Environmental Health.

Meyers, T. R., and Hendricks, J. D. "A Limited Epizootic of Neuroblastoma in Coho Salmon Reared in Chlorinated-Dechlorinated Water." *Journal of the National Cancer Institute* 72 (1984):299–310.

Crump, K. S., and Guess, H. A. "Drinking Water and Cancer: Review of Recent Epidemiologic Findings and Assessment of Risks." *Annual Review of Public Health* 3 (1982):339–57.

Gottlieb, M. S., and Carr, J. K. "Case-Control Cancer Mortality Study and Chlorination of Drinking Water in Louisiana." *Environmental Health Perspectives* 46 (1982):169–77.

Gottlieb, M. S., and Carr, J. K. "Cancer and Drinking Water in Louisiana: Colon and Rectum." *International Journal of Epidemiology* 10 (1981):117–25.

Simenhoff, M. L., et al. "Generation of Dimethylnitrosamine in Water Purification Systems: Detection in Human Blood Samples During Hemodialysis." *Journal of the American Medical Association* 250 (1983): 2020–24.

Dorson, W. J., Jr., et al. "Effects of Chloramines in Hemodialysis: A Symposium Report and Papers." *Contemporary Dialysis*, September 1984, pp. 29–41.

Singer, G. L. "The Need for Preventive Control of Drinking Water Contaminants." *Journal of the American Water Works Association* 74, no. 10 (1982):18–49.

Barrett, B. R. "Controlling the Entrance of Toxic Pollutants into U.S. Waters." *Environmental Science and Technology* 12 (1978):154–62.

Chapter 5: Vitamins and Minerals

Herbert, V. "The Vitamin Craze." *Archives of Internal Medicine* 140 (1980):173–76.

Periano, C., et al. "Enhancing Effects of Phenobarbitone and Butylated Hydroxytoluene on 2-Acetylaminofluorene-Induced Hepatogenic Tumorigenesis in the Rat." *Food and Chemical Toxicology* 15 (1977):93–96.

Environmental Protection Agency, Office of Toxic Substances. "Butylated Hydroxytoluene." Chemical Hazard Information Profile, Draft Report, August 13, 1984.

Witschi, H. D., et al. "Enhancement of Urethan Tumorigenesis in Mouse Lung by Butylated Hydroxytoluene." *Journal of the National Cancer Institute* 58 (1977):301–7.

Barclay, A. J. G., et al. "Vitamin A Supplements and Mortality Related to Measles: A Randomised Clinical Trial." *British Medical Journal* 294 (1987):294–96.

Levander, O. A. "Clinical Consequences of Low Selenium Intake and Its Relationship to Vitamin E." In *Vitamin E: Biochemical, Hematological and Clinical Aspects*, edited by Bertram Lublin and Lawrence J. Machlin, pp. 70–82. *Annals of the New York Academy of Sciences* 393 (1982).

Pauling, L., et al. "Incidence of Squamous-Cell Carcinoma in Hairless Mice Irradiated with Ultraviolet Light in Relation to Intake of Ascorbic Acid (Vitamin C) and of D,L-alpha-Tocopherol Acetate (Vitamin E)." In *Vitamic C: New Clinical Applications in Immunology, Lipid Metabolism and Cancer*, edited by A. Hanck, pp. 53–82. Bern, Stuttgart, Vienna: Hans Huber Publishers, 1982.

Basu, T. K. "Vitamin C–Aspirin Interactions." In ibid., pp. 83–90.

Hennekens, C. H. "Micronutrients and Cancer Prevention." *New England Journal of Medicine* 315 (1986):1288–89.

Menkes, M. S., et al. "Serum Beta-Carotene, Vitamins A and E, Selenium, and the Risk of Lung Cancer." *New England Journal of Medicine* 315 (1986):1250–54.

Cutler, R. G. "Antioxidants, Aging, and Longevity." Chap. 11 in *Free Radicals in Biology*, vol. 6, edited by W. A. Pryor. Orlando, Fla.: Academic Press, 1984.

Willet, W. C., et al. "Relation of Serum Vitamins A and E and Carotenoids to the Risk of Cancer." *New England Journal of Medicine* 310 (1984):430–34.

Basu, T. K., et al. "Plasma Vitamin A in Patients with Bronchial Carcinoma." *British Journal of Cancer* 33 (1976):119–21.

Panush, R. S., and Delafuente, J. C. "Vitamins and Immunocompetence." *World Review of Nutrition and Dietetics* 45 (1985):97–132.

Anderson, R. "The Immunostimulatory, Anti-inflammatory and Anti-allergic Properties of Ascorbate." *Advances in Nutrition Research* 6 (1984):19–45.

Miyama, T., et al. "Chronological Relationship Between Neurological Signs and Electrophysiological Changes in Rats with Methyl-

mercury Poisoning—Special Reference to Selenium Protection."
Archives of Toxicology 52 (1983):173–81.

Schmahl, D., and Eisenbrand, G. "Influence of Ascorbic Acid on
the Endogenous (Intragastral) Formation of N-Nitroso Compounds." In *Vitamic C: New Clinical Applications in Immunology,
Lipid Metabolism and Cancer*, edited by A. Hanck, pp. 91–102.
Bern, Stuttgart, Vienna: Hans Huber, 1982.

Cameron, E. "Vitamin C and Cancer; An Overview." In ibid., pp.
115–27.

Burton, G. W., et al. "Vitamin E as an Antioxidant *In Vitro and In
Vivo.*" In *Biology of Vitamin E: Ciba Foundation Symposium 101*,
pp. 4–18. London: Pitman Press, 1983.

Garland, C. F., and Garland, F. C. "Do Sunlight and Vitamin D
Reduce the Likelihood of Colon Cancer?" *International Journal of
Epidemiology* 9 (1980):227–31.

Garland, C., et al. "Dietary Vitamin D and Risk of Colorectal Cancer: 19-Year Prospective Study in Men." *Lancet* 1 (1985):307–9.

Shrive, W., and Landsbord, E. M., Jr. "Roles of Vitamins as Coen-
zymes." In *Nutrition and the Adult Micronutrients*, edited by
R. B. Alfin-Slater and D. Kritchevsky, pp. 1–71. New York: Plenum
Publishing, 1980.

Peto, R., et al. "Can Dietary Beta-Carotene Materially Reduce Human Cancer Rates?" *Nature* 290 (1981):201–8.

Wolf, G. "Is Dietary Beta-Carotene an Anti-Cancer Agent?" *Nutrition Reviews* 40, no. 9 (1982):257–61.

Lewin, S. *Vitamin C: Its Molecular Biology and Medical Potential.*
New York: Academic Press, 1976.

Gilman, A. G. Goodman, L. S., and Gilman, A. *The Pharmacolo-
gical Basis of Therapeutics.* New York: Macmillan Publishing, 1985.

Mettlin, C. S., et al. "Vitamin A and Lung Cancer." *Journal of the
National Cancer Institute* 62 (1979):1435–42.

Zeigler, R. G., et al. "Dietary Carotene and Vitamin A and Risk of
Lung Cancer Among White Men in New Jersey." *Journal of the
National Cancer Institute* 73 (1984):1429–35.

Medina, R., et al. "Selenium May Act as a Cancer Inhibitor." *Journal
of the American Medical Association* 246 (1981):1510.

Virtamo, J., et al. "Serum Selenium and the Risk of Coronary Heart
Disease and Stroke." *American Journal of Epidemiology* 122
(1985):276–82.

Clark, L. C. "The Epidemiology of Selenium and Cancer." *Federation Proceedings* 44 (1985):2584–89.

Salonen, J. T., et al. "Risk of Cancer in Relation to Serum Concentrations of Selenium and Vitamin A and E: Matched Case-Control Analysis of Prospective Data." *British Medical Journal*, 290 (1985):417–20.

Bjelke, E. "Dietary Vitamin A and Human Lung Cancer." *International Journal of Cancer* 15 (1975):561–65.

Levine, M. "New Concepts in the Biology and Biochemistry of Ascorbic Acid." *New England Journal of Medicine* 314 (1986):892–902.

Mettlin, C., et al. "Vitamin A and Lung Cancer." *Journal of the National Cancer Institute* 62 (1979):1435–38.

Whitting, L. A. "Vitamin E and Liquid Antioxidants." Chap. 9 in *Free Radicals in Biology*, edited by W. A. Pryor, vol. 4. Orlando, Fla.: Academic Press, 1980.

MacLennan, R., et al. 'Risk Factors for Lung Cancers in Singapore Chinese, A Population with High Female Incidence Rates." *International Journal of Cancer* 20 (1976):854–60.

Butterworth, C. E., Jr., et al. "Improvement in Cervical Dysplasia Associated with Folic Acid Therapy in Users of Oral Contraceptives." *American Journal of Clinical Nutrition 35* (1982):73–82.

Tannenbaum, A., and Silverstone, H. "The Genesis and Growth of Tumors: Effects of Varying the Level of B Vitamins in the Diet." *Cancer Research* 12 (1952):733–49.

Wattenberg, I. W. "Inhibition of Carcinogenic and Toxic Effects of Polycyclic Hydrocarbons by Phenolic Antioxidants and Ethoxyquin." *Journal of the National Cancer Institute* 48 (1972):1425–30.

Taylor, J. C. "Antioxidants and Emphysema." *City of Hope Quarterly* 14, no. 1 (1985):7–10.

Cook, M. G., and McNamara, P. "Effect of Dietary Vitamin E on Dimethylhydrazine-Induced Colonic Cancers in Mice." *Cancer Research* 40 (1980):1329–31.

Shute, W. E. *Complete . . . Updated Vitamin E Book*. New Canaan, Conn.: Keats Publishing, 1975.

Goldstein, I. M., et al. "Ceruloplasmin: A Scavenger of Superoxide Anion Radicals." *Journal of Biological and Chemistry* 254 (1979):4040–45.

Theron, A., and Anderson, R. "Investigation of the Protective Effects of the Antioxidants Ascorbate, Cysteine, and Dapsone on the Phagocyte-Mediated Oxidative Inactivation of Human Alpha-1-protease Inhibitor *In Vitro*." *American Review of Respiratory Disease* 132 (1985):1049–54.

Harpey, J. P., et al. "Homocystinuria Caused by 5,10-Methylene-tetrahydrofolate Reductase Deficiency: A Case in an Infant Responding to Methionine, Folinic Acid, Pyridoxine, and Vitamin B-12 Therapy." *Journal of Pediatrics* 2 (1981):275–78.

Committee on Recommended Dietary Allowances. *Recommended Dietary Allowances*, 9th ed. Washington, D.C.: Food and Nutrition Board, National Research Council, 1980.

Braganza, J. M. "Selenium Deficiency, Cystic Fibrosis, and Pancreatic Cancer." *Lancet* 2 (1985):1238.

Chapter 6: Exercise and Stress, Vegetables and Fruit

U.S. Congress. Senate. *Diet Related to Killer Diseases*. Vol. 2, *Obesity*. Hearings Before the Select Committee on Nutrition and Human Needs, March 24, 1977. 95th Cong., 1st sess.

*Goldblatt, P. B., et al. "Social Factors in Obesity." *Journal of the American Medical Association* 192 (1965):1039–44.

*Van Itallie, T. B. "Health Implications of Overweight and Obesity in the United States." *Annals of Internal Medicine* 103 (1985):938–42.

Kempner, W., et al. "Treatment of Massive Obesity with Rice/Reduction Diet Program." *Archives of Internal Medicine* 135 (1975):1575–84.

Kempner, W. "Treatment of Hypertensive Vascular Disease with a Rice Diet." *American Journal of Medicine* 4 (1948):545–52.

Marks, H. H. "Influence of Obesity on Morbidity and Mortality." *Bulletin of New York Academy of Medicine* 36 (1960):292–312.

National Institutes of Health Consensus Development Panel on the Health Implications of Obesity. Conference Statement. *Annals of Internal Medicine* 103 (1984):1073–77.

Brown, J. M., et al. "Cardiac Complications of Protein-Sparing Modified Fasting." *Journal of the American Medical Association* 240 (1978):120–22.

*Roberts, H. J. "The Hazards of Very-Low-Calorie Dieting." *American Journal of Clinical Nutrition* 41 (1985):171–72.

Lantigua, R. A., et al. "Cardiac Arrhythmias Associated with a Liquid Protein Diet for the Treatment of Obesity." *New England Journal of Medicine* 303 (1980):735–38.

Van Itallie, T. B., and Yang, M. U. "Cardiac Dysfunction in Obese Dieters: A Potentially Lethal Complication of Rapid, Massive Weight Loss." *American Journal of Clinical Nutrition* 39 (1984):695–70.

Isner, J. M., et al. "Sudden Unexpected Death in Avid Dieters Using the Liquid-Protein-Modified-Fast Diet: Observation in 17 Patients and the Role of the Prolonged QT Interval." *Circulation* 60 (1979):1401–12.

Evans, W. J., and Hughes, V. A. "Dietary Carbohydrates and Endurance Exercise." *American Journal of Clinical Nutrition* 41 (1985):1146–54.

Astrand, P. O. "Something Old, Something New." *Nutrition Today* 3 (1968):9–11.

Burfoot, Amby. "The Vegetarian Diet for Athletes." *Runner's World*, February 1978.

Lamb, J. "The Healthiest Diet." *Sports Illustrated* 46 (1977):68–72.

Hartung, H. G., et al. "Relation of Diet to High Density Lipoprotein Cholesterol in Middle-Aged Marathon Runners, Joggers, and Inactive Men." *New England Journal of Medicine* 302 (1980):357–63.

Cantwell, J. D., and Fletcher, G. F. "Cardiac Complications While Jogging." *Journal of the American Medical Association* 210 (1969):130–31.

Noakes, T., et al. "Coronary Heart Disease in Marathon Runners." *Annals of the New York Academy of Science* 301 (1977):593–619.

Caseleth, B. R., et al. "Psychosocial Correlates of Survival in Advanced Malignant Disease?" *New England Journal of Medicine* 312 (1985):1570–72.

Cousins, N. "Stress." *Journal of the American Medical Association* 242 (1979):459.

Cousins, N. *Anatomy of an Illness*. New York: W. W. Norton & Co., 1979.

Rosch, P. J. "Stress and Illness." *Journal of the American Medical Association* 242 (1979):427–28.

Selye, H. *Stress Without Disease*. Philadelphia: J. B. Lippincott, 1974.

Selye, H. *The Stress of Life*. New York: McGraw-Hill, 1976.

Rainsford, G. L., and Schuman, S. H. "The Family in Crisis: A Case Study of Overwhelming Illness and Stress." *Journal of the American Medical Association* 246 (1981):60–63.

Anderson, James W. "Health Implications of Wheat Fiber." *American Journal of Clinical Nutrition* 41 (1985):1103–12.

Storch, K., et al. "Oat Bran Muffins Lower Serum Cholesterol of Healthy Young Men." *American Journal of Clinical Nutrition* 41 (1985):846.

Anderson, James W., et al. "Hypocholesterolemic Effects of Oat-

Bran or Bean Intake for Hypercholesterolemic Men." *American Journal of Clinical Nutrition* 40 (1984):1146–55.

Anderson, James W., et al. "Hypolipidemic Effects of High-Carbohydrate, High-Fiber Diets." *Metabolism* 29 (1980):551–58.

Chapter 8: Eliminating Processed Foods

Food Additives and the Consumer. London: Commission of the European Community, 1980.

National Academy of Sciences Research Council. "Food Colors: The Use of Chemical Additives in Food Processing." Washington, D.C.: National Academy of Sciences, 1971.

Code of Federal Regulations. Title 21, part 181–184 (GRAS list). Washington, D.C.: Office of the Federal Register, July 1985.

*Marine, G., and Van Allen, J. *Food Pollution*. New York: Holt, Rinehart & Winston, 1972.

Burton, E. *The Pageant of Early Victorian England, 1837–1861*. New York: Charles Scribner's Sons, 1972.

Lijinsky, W., and Shubik, P. "Benzo[a]pyrene and Other Polynuclear Hydrocarbons in Charcoal-Broiled Meats." *Science* 145 (1964):53–61.

Levine, A. S., et al. "Food Technology: A Primer for Physicians." *New England Journal of Medicine* 312 (1985):628–34.

Williams. R. J. "Should the Science-Based Food Industry Be Expected to Advance?" Lecture before the National Academy of Science, October 21, 1970.

"Carcinogenic Effect Seen in Preliminary Report on FD&C Yellow #6." *Food Chemistry News*, no. 19 (October 1, 1984).

Lehmann, P. "More Than You Ever Thought You Would Know About Food Additives." *FDA Consumer*, HHS Publication No. 82-2160.

Hargraves, W. A., and Pariza, M. W. "Mutagens in Cooked Foods." *Journal of Environmental Sciences and Health, Part C: Environmental Carcinogenisis Review* C2, 1 (1984):1–49.

Hartley, D. E. "Sulfiting Agents in Food." *Journal of Environmental Health* 47 (1984):1.

Weston, R. J. "Analysis of Cereals, Malted Foods and Dried Legumes for N-Nitrosodimethylamines." *Journal of the Science of Food and Agriculture* 35 (1984):782–86.

Krone, C. A., and Iwaoka, W. T. "Occurrence of Mutagens in Canned Foods." *Mutation Research* 141 (1984):131–34.

Munro, E. C., et al. "A Carcinogenicity Study of Commercial Saccharin in the Rat." *Toxicology and Applied Pharmacology* 32 (1975):513–17.

Olney, J. W. "Another View of Aspartame." In *Sweeteners: Issues and Uncertainties,* National Academy of Sciences Forum, Washington, D.C., 1975.

Roitman, J. N. "Ingestion of Pyrrolizidine Alkaloids: A Health Hazard of Global Proportions." Chap. 22 in *Xenobiotics in Food and Feed,* edited by John W. Finley and Daniel E. Schwarz. Series 234, ACS Symposium. Wasington, D.C.: American Chemical Society, 1983.

Krone, C. A., et al. "Mutagen Formation in Processed Foods." Chapt. 7 in ibid.

Herikstad, H. "Screening for Mutagens in Norwegian Foods by the Ames Test." *Journal of the Science of Food and Agriculture* 35 (1984):900–908.

Williams, R. *Nutrition Against Disease.* New York, London: Pitman Publishing Co., 1971.

Johnson, K. "How to Read a Food Label." *EastWest Journal* 15, no. 1 (1985):61–66.

Thomas, B., et al. "Lead and Cadmium Content of Some Canned Fruit and Vegetables." *Journal of the Science of Food and Agriculture* 24 (1973):447–49.

Hansen, P. D. "Lead and Arsenic Levels in Wine Produced in Vineyards in Australia." *Journal of the Science of Food and Agriculture* 35 (1984):215–20.

Chapter 9: Minimizing Environmental Hazards

Bardana, E. J., and Montanaro, A. "Tight Building Syndrome." *Immunology and Allergy Practice* 8, no. 3 (1986):74–88.

County of Los Angeles, Department of Health Services, Hazardous Waste Control Program. "Toxic Chemicals in My Home? Absolutely!" July 1985.

County of Los Angeles, Department of Health Services. "Polymer-Fume Fever." *Public Health Letter* 7, no. 8 (August 1985):2.

Harris, J. C., et al. "Toxicology of Urea Formaldehyde and Polyurethane Foam Insulation." *Journal of the American Medical Association* 245 (1981):243–46.

Vaisrub, S. "The Health Cost of Tight Homes." *Journal of the American Medical Association* 245 (1981):267–68.

Meier, B., and Trost, C. "Doctor Cites Benzene Exposure as Likely Cause of High Cancer Death Rate at a Shell Refinery." *Wall Street Journal*, February 14, 1985.

Fritsch, A. J., ed., *The Household Pollutants Guide*. New York: Anchor Books/Center for Science in the Public interest, 1978.

*Raloff, J. "Agent Orange: What Isn't Settled." *Science News* 125, no. 20 (1984):314–17.

Meyerhoff, Al. "Pesticides: Food and Water May Be Hazardous to Health." *Los Angeles Times*, June 26, 1985.

*Mossing, M. L., et al. "Organochlorine Pesticides in Blood of Persons from El Paso, Texas." *Journal of Environmental Health* 47 (1984):312–13.

Infante, P. F., and Schneiderman, M. A. "Formaldehyde, Lung Cancer and Bronchitis." *Lancet* 1 (1986):436–37.

Acheson, E. D., et al. "Formaldehyde Process Workers and Lung Cancer." *Lancet* 1 (1984):1066–69.

Selikoff, I. J. "Twenty Lessons from Asbestos: A Bitter Harvest of Scientific Information." *Journal of Environmental Health* 47 (1984):140–44.

Tartaglia, M., et al. "Radon and Its Progeny in the Indoor Environment." *Journal of Environmental Health* 47 (1984):62–67.

Nylander, M. "Mercury in Pituitary Glands of Dentists." *Lancet* 1 (1986):442.

Cooley, R. L., and Barkmeyer, W. W. "Mercury Vapor Emitted During Ultraspeed Cutting of Amalgam." *Indiana State Dental Association Journal* 57 (1978):28–31.

Shapiro, I. M., et al. "Neurophysiological and Neuropsychological Function in Mercury-Exposed Dentists." *Lancet* 1 (1982):1147–50.

Hallenbeck, W. H. "Human Health Effects of Exposure to Cadmium." *Experientia* 2 (1984):136–42.

Tsubaki, T., et al. *Minimata Disease: Methylmercury Poisoning in Minimata and Niigata, Japan*. Amsterdam, New York, Oxford: Elsevier, 1977.

Calvan, R. A. "Growing Concerns About Gasoline Vapors." *EPA Journal* 10, no. 7 (1981):20–21.

Ruckelshaus, W. D. "Dealing with EDB, a Dangerous Pesticide." *EPA Journal* 10, no. 1 (1981):16–17.

Environmental Protection Agency. "EDB Facts." Public Affairs Bulletin A-107, February 3, 1984.

Silverberg, E. "Cancer Statistics, 1985." *Ca—A Cancer Journal of Clinicians* 35 (1985):19–35; 36 (1986):9–25.

Chapter 10: Prescription and Nonprescription Drugs

Kaufman, J., et al. *Over-the-Counter Pills That Don't Work.* New York: Pantheon Books/Public Citizen Health Research Group, 1983.

Wolfe, S. M., et al. *Pills That Don't Work.* New York: Farrar, Straus & Giroux/Ralph Nader's Health Research Group, 1980.

Hoover, R., and Fraumeni, J. F. "Drug-Induced Cancer." *Cancer* 47 (1981):1071–80.

Paffenberger, R. S., et al. "Cancer Risk as Related to Use of Oral Contraceptives During Fertile Years." *Cancer Research* 39 (1977):1887–95.

Krantz, J. C., Jr., and White, J. M. "Aspirin: Four Score and More." *Southern Medical Journal* 73 (1980):1630–34.

*Barbeau, A., et al. "Ecogenetics of Parkinson's Disease: 4-Hydroxylation of Debrisoquine." *Lancet* 2 (1985):1213–16.

*Davis, G. C., et al. "Chronic Parkinsonism Secondary to Intravenous Injection of Meperidine Analogues." *Psychiatric Research* 1 (1979):249–54.

*Langston, J. W., and Ballard, P. A. "Parkinson's Disease in a Chemist Working with 1-Methyl-4-phenyl-1,2,5,6-tetrahydropyridine." *New England Journal of Medicine* 309 (1983):310–13.

*Norvenius, G., et al. "Phenylpropanolamine and Mental Disturbances." *Lancet* 2 (1979):1367–68.

Mueller, S. M. "Phenylpropanolamine, a Non-Prescription Drug with Potentially Fatal Side Effects." *New England Journal of Medicine* 308 (1983):653.

*Koch-Weser, J. et al. "Drug Therapy: Drug Disposition in Old Age." *New England Journal of Medicine* 306 (1982):1081–88.

*McDermott, M. T. et al. "Radioiodine-Induced Thyroid Storm: Case Report and Literature Review." *American Journal of Medicine* 75 (1983):353–59.

Roxe, D. M. "Toxic Nephropathy from Diagnostic and Therapeutic Agents." *American Journal of Medicine* 69 (1980):759–66.

Zafrani, E. S. et al. "Drug-Induced Vascular Lesions of the Liver." *Archives of Internal Medicine* 143 (1983):495–502.

Venning, G. R. "Identification of Adverse Reactions to New Drugs: What Have Been the Important Adverse Reactions Since Thalidomide?" *British Medical Journal* 286 (1983):199–202.

Piper, J., et al. "Heavy Phenacitin Use and Bladder Cancer in Women Aged 20 to 49 Years." *New England Journal of Medicine* 313 (1985):292–95.

Reidenberg, M. M., et al. "Lupus Erythematosus-like Disease Due to Hydrazine." *American Journal of Medicine* 75 (1983):365–70.

Toth, B. "Synthetic and Naturally Occurring Hydrazines as Possible Cancer Causative Agents." *Cancer Research* 35 (1975):3693–97.

Zierlier, S., and Rothman, K. J. "Congenital Heart Disease in Relation to Maternal Use of Bendectin and Other Drugs in Early Pregnancy." *New England Journal of Medicine* 313 (1985):347–52.

Larkin, T. "Cortisone: The Limits of a Miracle." *FDA Consumer* 19, no. 7 (1985):26–29.

Chapters 11 & 12: Eliminating Fish, Poultry, Meats, and Fats

Weight Watchers Personal Program. Weight Watchers International, 1980.

Fleming, D. W., et al. "Pasteurized Milk as a Vehicle of Infection in an Outbreak of Listeriosis." *New England Journal of Medicine* 312 (1985):404–7.

Margolin, S. "Mercury in Marine Seafood: The Scientific Medical Margin of Safety as a Guide to the Potential Risk to Public Health." *World Review of Nutrition and Dietetics* 34 (1980):182–265.

Sawyer, Dianne. "A Report on Presence of Salmonella Bacteria in Processed, Inspected Poultry in the United States." *60 Minutes*, CBS Television News Service, March 29, 1987.

Rouse, I. L., et al. "Vegetarian Diet, Blood Pressure and Cardiovascular Risk." *Australian and New Zealand Journal of Medicine* 14 (1984):439–43.

Friend, B. "Nutrients in U.S. Food Supply: A Review of Trends." *American Journal of Clinical Nutrition* 20 (1967):907.

West, R. O., and Hayes, O. B. "Diet and Serum Cholesterol Levels: A Comparison Between Vegetarians and Nonvegetarians in a Seventh-Day Adventist Group." *American Journal of Clinial Nutrition* 21 (1969):853–62.

Goldberg, M. J., et al. "Comparison of the Fecal Microflora of Seventh-Day Adventists with Individuals Consuming a General Diet: Implications Concerning Colonic Carcinoma." *Annals of Surgery* 186 (1977):97–100.

Akers, Keith. *A Vegetarian Sourcebook.* New York: G. P. Putnam's Sons, 1983.

Herzog, D. B., and Copeland, P. M. "Eating Disorders." *New England Journal of Medicine* 313 (1985):295–303.

Harper, A. E. "Origin of Recommended Dietary Allowances: An

Historic Overview." *American Journal of Clinical Nutrition* 41 (1985):140–48.

Blumenfeld, A. *Heart Attack: Are You a Candidate?* New York: Paul Eriksson, 1964.

Corwin, Miles. "Where Does All That Grease Go?" *Los Angeles Times*, December 29, 1984.

Snapper, I. *Chinese Lessons to Western Medicine.* New York: Interscience Publications, 1941.

Illingworth, D. R., et al. "Inhibition of Low Density Lipoprotein Synthesis by Dietary Omega-3 Fatty Acids in Humans." *Arteriosclerosis* 4 (1984):270–75.

Lee, T. H., et al. "Effect of Dietary Enrichment with Eicosapentaenoic and Docosahexaenoic Acids on *In Vitro* Neutrophil and Monocyte Leukotriene Generation and Neutrophil Function." *New England Journal of Medicine* 312 (1985):1217–24.

Phillipson, B. E., et al. "Reduction of Plasma Lipids, Lipoproteins and Apoproteins by Dietary Fish Oils in Patients with Hypertriglyceridemia." *New England Journal of Medicine* 312 (1985): 1210–16.

Norris, P. G., et al. "Effect of Dietary Supplementation with Fish Oil on Systolic Blood Pressure in Mild Essential Hypertension." *British Medical Journal* 293 (1986):104–5.

Kromhout, D., et al. "The Inverse Relation Between Fish Consumption and 20-Year Mortality from Coronary Heart Disease." *New England Journal of Medicine* 312 (1985):1205–9.

Cruse, P., et al. "Cholesterol and Colon Cancer." *Lancet* 2 (1979):43–44.

Kinlen, L. J. "Meat and Fat Consumption and Cancer Mortality: A Study of Strict Religious Orders in Britain." *Lancet* 1 (1982): 946–49.

Wynder, E. L., et al. "Cancer and Coronary Artery Disease Among Seventh-Day Adventists." *Cancer* (1959):1016–28.

Part III: THE RESEARCH

Chapter 16: The Myths of the Modern Diseases

Baas, J. H., and Handerson, H. E. *Outlines of the History of Medicine.* Huntington, N.Y.: Robert E. Krieger Publishing, 1889.

Sigerist, H. E. *A History of Medicine.* Vol. 1, *Primitive and Archaic Medicine.* New York: Oxford University Press, 1951.

Nriagu, J. O. "Saturnine Gout Among Roman Aristocrats: Did Lead Poisoning Contribute to the Downfall of the Roman Empire?" *New England Journal of Medicine* 308 (1981):660–63.

Infectious Diseases

Kass, E. H. "Infectious Diseases and Social Change." *Journal of Infectious Diseases* 123 (1971):110–14.

Coronary Heart Disease

Parrish, H. M. "Has Mankind Always Had Coronary Heart Disease?" *Journal of the Indiana Medical Association* 55 (1962):464–71.

Leibowitz, J. O. *History of Coronary Heart Disease*. Berkeley, Los Angeles: University of California Press, 1970.

Sprague, Howard B. "Environment in Relation to Coronary Artery Disease." *Archives of Environmental Health* 13 (1966):4–12.

Michaels, L. "Aetiology of Coronary Artery Disease: An Historical Approach." *British Heart Journal* 28 (1966):258–64.

Ruffer, M. A. "On Arterial Lesions Found in Egyptian Mummies." *Journal of Pathology and Bacteriology* 15 (1911):453–59.

Ruffer, M. A. *Studies in the Paleopathology of Egypt*. Chicago: University of Chicago Press, 1921.

Shattock, A. "A Report on the Pathological Condition of the Aorta of King Meneptah." *Proceedings of the Royal Society of Medicine* 2 (1909):122.

Long, A. R. "Cardiovascular Renal Disease: Report on a Case Three Thousand Years Ago." *Archives of Pathology* (Chicago) 12 (1931):92–94.

Zimmerman, M. R. "The Mummies of the Tomb of Nebwenenef: Paleopathology and Archaeology." *Jarce XIV*. J. J. Augustin, Locust Valley, NY, pp. 33–36, 1977.

Hunter, J. *A Treatise on blood, inflammation and gun-shot wounds to which is prefixed, a short account of the author's life by his brother-in-law Everard Home*. London: G. Nicol, 1794.

Heberden, W. "Some account of a disorder of the breast." *Medical Transactions, Royal College of Physicians, London* 2 (1772):59–67.

Heberden, W. "A letter to Dr. Heberden, concerning the angina pectoris: and Dr. Heberden's account of the dissection of one, who had been troubled with the disorder." *Medical Transactions, Royal College of Physicians, London* 3 (1785):1–11.

Herrick, J. B. "Clinical Features of Sudden Obstruction of the Coronary Arteries." *Journal of the American Medical Association* 59 (1912):2015–20.

Herrick, J. B. *A Short History of Cardiology.* Springfield, Ill.: Thomas & Co., 1942.

White, Paul Dudley. *My Life in Medicine: An Autobiographical Memoir.* Boston: Gambit & Co., 1971.

Mead, Melinda. "Cardiovascular Mortality in the South-Eastern U.S.: The Coastal Plains Enigma." *Social Science and Medicine, Medical Geography,* 13 (1985):257–65.

Keys, Ancel, et al. *Seven Countries.* Cambridge, Mass.: Harvard University Press, 1980.

Brown, M. S., and Goldstein, J. L. "How LDL Receptors Influence Cholesterol and Atherosclerosis." *Scientific American* 251 (1984):58–66.

Turpeinen, O. "Effects of Cholesterol-Lowering Diet on Coronary Heart Disease and Other Causes." *Circulation* 59 (1979):1–7.

Racline, L., et al. "Atherosclerosis: Current Concepts." *American Journal of Surgery* 141 (1981):638–43.

*Kagawa, Y. "Impact of Westernization on the Nutrition of Japanese: Changes in Physique, Cancer, Longevity and Centenarians." *Preventive Medicine* 7 (1978):205–17.

Kimura, N. "Changing Patterns of Coronary Heart Disease, Stroke and Nutrient Intake in Japan." *Preventive Medicine* 12 (1983):37–39.

*Dreyfuss, F. "The Incidence of Myocardial Infarction in Various Communities In Israel." *Amerian Heart Journal* 45 (1953):749–55.

*Toor, M., et al. "Serum Lipids and Atherosclerosis Among Yemenite Immigrants in Israel." *Lancet* 1 (1957):1270–73.

*Mabuchi, K., and Maruchi, N. "The Major Causes of Death in the United States and Japan." *Preventive Medicine* 1 (1972):252–54.

Robertson, D. L. "Epidemiologic Studies of Coronary Heart Disease and Stroke in Japanese Men in Japan, Hawaii and California: Coronary Heart Disease Risk in Japan and Hawaii." *American Journal of Cardiology* 34 (1977):244–49.

Welin, L., et al. "Why Is The Incidence of Ischemic Heart Disease in Sweden Increasing? Study of Men Born in 1913 and 1923." *Lancet* 1 (1983):1087–89.

298   *Bibliography*

Blankenhorn, D. H. "Will Atheroma Regress with Diet and Exercise?" *American Journal of Surgery* 141 (1981):644–45.

Garcia-Palmieri, M. R., et al. "Increased Physical Activity: A Protective Factor Against Heart Attacks in Puerto Rico." *American Journal of Cardiology* 50 (1982):749–55.

Insull, W., et al. "The Lipid Research Clinics Coronary Primary Prevention Trial Results: I. Reduction in Incidence of Coronary Heart Disease." *Journal of the American Medical Association* 251 (1984):351–64.

Insull, W., et al. "The Lipid Research Clinics Coronary Primary Prevention Trial Results: II. The Relationship of Reduction in Incidence of Coronary Heart Disease to Cholesterol Lowering." *Journal of the American Medical Association* 251 (1984):365–74.

Berenson, G. S., and Epstein, F. H. "Conference on Blood Lipids in Children: Optimal Levels for Early Prevention of Coronary Artery Disease. Workshop Report." *Preventive Medicine* 12 (1983):6.

Kaplan, N., and Stamler, J. *Prevention of Coronary Heart Disease.* New York: W. B. Saunders & Co., 1983.

Lanen, R. M., and Shekell J., eds. *Childhood Prevention of Atherosclerosis and Hypertension.* New York: Raven Press, 1980.

Arntzenius, A. C., et al. "Diet, Lipoproteins and the Progression of Coronary Atherosclerosis: The Leiden Intervention Trial." *New England Journal of Medicine* 312 (1985):805–11.

Jacobsen, N. L. "The Controversy over the Relationship of Animal Fats to Heart Disease." *BioScience*, Journal Paper No. J-7516 of the Iowa Agriculture and Home Economics Experimental Station, Ames Project 1910 of Iowa State University, Ames, Iowa.

Gsell, D., and Mayer, J. "Low Blood Cholesterol Associated with High Calorie, High Saturated Fat Intake in a Swiss Alpine Village Population." *American Journal of Clinical Nutrition* 10 (1962):471–79.

Castelli, W. P., et al. "HDL Cholesterol and Other Lipids in Coronary Heart Disease: The Cooperative Lipoprotein Phenotyping Study." *Circulation* 5 (1977):245–51.

Barnard, R. J., et al. "Effects of a High-Complex-Carbohydrate Diet and Daily Walking on Blood Pressure and Medication Status of Hypertensive Patients." *Journal of Cardiac Rehabilitation* 3 (1983):839–46.

McCarron, D. A., and Morris, C. D. "Blood Pressure Response to Oral Calcium in Persons with Mild to Moderate Hypertension: A

Randomized Double-Blind, Placebo-Controlled, Cross-Over Trial." *Annals of Internal Medicine* 103, no. 6, pt. 1 (1985):825–31.

Schroeder, H. A. "Cadmium as a Factor in Hypertension." *Journal of Chronic Disease* 18 (1965):647–56.

*Fye, W. B. "Acute Coronary Occlusion Always Results in Death— or Does It? The Observations of William T. Porter." *Circulation* 71 (1985):4–10.

Kushi, L. H., et al. "Diet and 20-Year Mortality from Coronary Heart Disease: The Ireland-Boston Diet-Heart Study." *New England Journal of Medicine* 312 (1985):811–18.

*Cooper, R. "Rising Death Rate in the Soviet Union: Impact of Coronary Heart Disease." *New England Journal of Medicine* 304 (1981):1259–65.

*Larsson, B., et al. "Abdominal Adipose Tissue Distribution, Obesity, and Risk of Cardiovascular Disease and Death: 13-Year Follow-Up of Participants in the Study of Men Born in 1913." *British Medical Journal* 288 (1984):1401–04.

Paffenberger, R. S., Jr., et al. "A Natural History of Athleticism and Cardiovascular Health." *Journal of the American Medical Association* 252 (1984):491–95.

Eisenberg, M. S., et al. "Sudden Cardiac Death." *Scientific American* 254 (1986):37–43.

Enos, J., et al. "Pathogenesis of Coronary Disease in American Soldiers Killed in Korea." *Journal of the American Medical Association* 158 (1955):912–14.

McNamara, J. J., et al. "Coronary Artery Disease in Vietnam Casualties." *Journal of the American Medical Association* 216 (1971):1185–92.

*"Leading Cause of Death in Japan." In *Vital Statistics*. Statistics and Information Department, Ministry of Health and Welfare, Japan.

*Keys, A., et al. "Lessons from Serum Cholesterol Studies in Japan, Hawaii, and Los Angeles." *Annals of Internal Medicine* 48 (1958):83–94.

Ueshima, H., et al. "Alcohol Intake and Hypertension Among Urban and Rural Japanese Population." *Journal of Chronic Diseases* 37 (1984):585–92.

Thulin, Thomas, et al. "Chronic Alcoholism and Hypertension." *Lancet* 1 (1984):168–69.

Multiple Risk Factor Intervention Trial Research Group. "Multiple Risk Factor Intervention Trial: Risk Factor Change and Mortality

Results." *Journal of the American Medical Association* 248 (1980):1467–77.

Sinclair, H. "Eicosapentaenoic Acid and Ischemic Heart Disease." *Lancet* 2 (1982):393.

Verlangieri, A. J., et al. "Fruit and Vegetable Consumption and Cardiovascular Mortality." *Medical Hypotheses* 16 (1985):7–15.

Acheson, R. M., and Williams, D. R. R. "Does Consumption of Fruits and Vegetables Protect Against Stroke?" *Lancet* 1 (1983):1191–93.

Cancer

Newman, W. "Notes from Registers of Market Deeping, 1711–1723." *British Medical Journal* 1 (1896):915–18.

Williams, W. R. *The Natural History of Cancer.* New York: William Wood & Co., 1908.

Hoffman, F. L. *The Mortality from Cancer Throughout the World.* New York: Prudential Press, 1915.

Taylor, H. C. *Cancer: Its Study and Prevention.* Philadelphia and New York: Lea & Febiger, 1915.

Carson, Rachel. *Silent Spring.* New York: Fawcett Crest, 1962.

Bailer, J. C., III, and Smith, E. M. "Progress Against Cancer?" *New England Journal of Medicine* 314 (1986):1226–32.

Silverberg, E. "Cancer Statistics, 1985." *Ca—A Cancer Journal for Clinicians* 35 (1985):19–35; 36 (1986):9–25.

Fraumeni, J. F., Jr. *An Approach to Cancer Etiology and Control.* New York: Academic Press, 1975.

Wynder, E. L., and Gori, G. B. "Contribution of the Environment to Cancer Incidence: An Epidemiologic Exercise." *Journal of the National Cancer Institute* 58 (1977):825–32.

Doll, R., and Peto, R. "The Causes of Cancer: Quantitative Estimates of Avoidable Risks of Cancer in the United States Today." *Journal of the National Cancer Institute* 66 (1981):1245–48.

Task Force on Environmental Cancer and Heart and Lung Disease. "Environmental Cancer and Heart and Lung Disease." Fifth Annual Report to Congress, 1982; Sixth Annual Report to Congress, 1983.

Kolonel, L. N., et al. "Role of Diet in Cancer Incidence in Hawaii." *Cancer Research* 43 (1983):2397s–2402s.

Stemmermann, G. N., et al. "Serum Cholesterol and Colon Cancer

Incidence in Hawaiian Japanese Men." *Journal of the National Cancer Institute* 67 (1981):1179–82.

Carroll, K. K., et al. "Dietary Fat and Mammary Cancer." *Canadian Medical Association Journal* 98 (1971):590–98.

Carroll, K. K. "Essential Fatty Acids in Relation to Mammary Carcinogenesis." *Progress in Lipid Research* 20 (1981):685–90.

Cohen, Leonard A. "Diet and Cancer." *Scientific American* 257 (1987):44–48.

Törnberg, S. A., et al. "Risks of Cancer of the Colon and Rectum in Relation to Serum Cholesterol and Beta-Lipoprotein." *New England Journal of Medicine* 315 (1987):1629–33.

Phillips, R. L. Cancer Among Seventh-Day Adventists." *Journal of Environmental Pathology and Toxicology* 3 (1980):157–69.

Miller, R. W. "Carcinogens in Drinking Water." *Pediatrics* 57 (1976): 462–64.

Maugh, T. N. "New Study Links Chlorination and Cancer." *Science* 211 (1981):694.

Hems, G. "The Contribution of Diet and Childbearing to Breast-Cancer Rates." *British Journal of Cancer* 37 (1978):974–82.

Snowden, D. A., and Phillips, R. L. "Coffee Consumption and Risk of Fatal Cancers." *American Journal of Public Health* 74 (1984):820–23.

Gershbein, L. L., et al. "Influence of Stress on Lesion Growth and on Survival of Animals Bearing Parenteral and Intracerebral Leukemia L-1210 and Walker Tumors." *Oncology* 30 (1974):429–36.

Pearce, M. L., and Dayton, S. "Incidence of Cancer in Men on a Diet High in Polyunsaturated Fat." *Lancet* 1 (1971):464–67.

Visintainer, M. A., et al. "Tumor Rejection in Rats After Inescapable or Escapable Shock." *Science* 216 (1982):437–41.

Snowden, D. A., et al. "Diet, Obesity and Risk of Fatal Prostate Cancer." *American Journal of Epidemiology* 120 (1984):244–50.

Tuyns, A. J., et al. "Nutrition, Alcohol, Esophageal Cancer." *Bulletin of Cancer* (Paris) 65 (1978):59–64.

Simone, C. B. "How High Is Your Cancer Risk?" Talk delivered to the American Academy of Allergy Annual Conference, February 1984.

Edmondson, H. A., et al. "Liver-Cell Adenomas Associated with Use of Oral Contraceptives." *New England Journal of Medicine* 294 (1976):470.

Graham, S., et al. "Diet in the Epidemiology of Breast Cancer." *American Journal of Epidemiology* 116 (1982):68–75.

Reddy, B. S., et al. "Diet and Metabolism: Large Bowel Cancer." *Cancer* 39 (1977):1815–1819.

Reddy, B. S., et al. "Nutrition and Its Relationship to Cancer." *Advances in Cancer Research* 32 (1980):237–44.

Lowenfels, A. B., and Anderson, M. E. "Diet and Cancer." *Cancer* 39 (1977):1809–14.

Gori, G. B. "Food as a Factor in the Etiology of Certain Human Cancers." *Food Technology* 33, no. 12 (1979):48–56.

Alpert, M. E., et al. "Association Between Aflatoxin Content of Food and Hepatoma Frequency in Uganda." *Cancer* 28 (1971):253–56.

Blondell, J. M. "The Anti-Carcinogenic Effect of Magnesium." *Medical Hypotheses* 8 (1980):863–71.

Shils, M. E. "Diet and Nutrition as Modifying Factors in Tumor Development." *Medical Clinics of North America* 63 (1979):1027–41.

Ernster, V. L., et al. "Effects of Caffeine-Free Diet on Benign Breast Disease: A Randomized Trial." *Surgery* 91 (1982):263–67.

Marshall, J., et al. "Diet in the Epidemiology of Oral Cancer." *Nutrition and Cancer* 3 (1982):145–49.

Diabetes

Osler, Sir William. *The Principles and Practice of Medicine*. 2d ed. New York: D. Appleton & Co., 1895.

American Diabetes Association. "1985 Fact Sheet On Diabetes."

O'Dea, K. "Marked Improvement in Carbohydrate and Lipid Metabolism in Diabetic Australian Aborigines After Temporary Reversion to Traditional Lifestyle." *Diabetes* 33 (1984):596–603.

Anderson, James W., and Gustafson, Nancy J. "Type II Diabetes: Current Nutrition Management Concepts." *Geriatrics* 41, no. 8 (1986):28–33.

Alzheimer's Disease

Kokmen, E. "Dementia—Alzheimer Type." *Mayo Clinic Proceedings* 59 (1984):35–42.

Beck, J. C., et al. "Dementia in the Elderly: The Silent Epidemic." *Annals of Internal Medicine* 97 (1982):231–41.

Corkin, S., et al. *Aging*. Vol. 19, *Alzheimer's Disease: A Report of Progress in Research*. New York: Raven Press, 1982.

Katzman, R. "The Prevalence and Malignancy of Alzheimer's Disease." *Archives of Neurology* 33 (1976):217–18.

Mace, N., and Rabins, P. *The 36-Hour Day.* Baltimore: Johns Hopkins University Press, 1981.

*Alzheimer, A. "Über eine eigenartige Erkrankung der Hirnrinde." *Allgemeine Zeitschrift für Psychiatrie* 64 (1907):146–48.

Clark, Matt, with Gosnell, Mariana, et al. "A Slow Death of the Mind." *Newsweek*, December 3, 1984.

Delabar, J-M., et al. "Beta Amyloid Gene Duplication in Alzheimer's Disease and Karyotypically Normal Down Syndrome." *Science* 235 (1987):1390–92.

"A Possible Diagnostic Test for Alzheimer's?" *Science* 234 (1986):1324.

St. George-Hyslop, Peter H., et al. "Absence of Duplication of Chromosome 21 Genes in Familial and Sporadic Alzheimer's Disease." *Science* 238 (1987):664–66.

Tanzi, Rudolph E., et al. "The Amyloid β Protein Gene Is Not Duplicated in Brains from Patients with Alzheimer's Disease." *Science* 238 (1987):666–69.

Podlisny, Marcia Berman, et al. "Gene Dosage of the Amyloid β Precursor Protein in Alzheimer's Disease." *Science* 238 (1987):669–71.

Multiple Sclerosis

National Multiple Sclerosis Society. "Fact Sheet on MS Statistics."

Swank, R. L., and Pullen, M. H. *The Multiple Sclerosis Diet Book.* Beaverton, Ore.: Willamette Publishing Co., 1972.

Limburg, C. C. "Geographical Distribution of Multiple Sclerosis and Its Estimated Prevalence in the United States." *Archives of Nervous and Mental Disease* 25 (1950):15–20.

Carswell, Robert. "Pathological Anatomy." In *Illustrations of the Elementary Forms of Disease.* London: Longmans, 1838.

Cruveilhier, Jean. *Anatomie pathologique du corps humain ou déscriptions avec figures lithographiées et caloriées des diverse alterations morbides dont le corps humain est susceptible.* Vol 2. Paris: J. B. Ballière, 1842.

Amyotrophic Lateral Sclerosis

Charcot, Jean-Martin. "Histologie de la sclerose en plaques." *Gazette d'Hôpital* 41 (1868):554–57, 566.

Amyotrophic Lateral Sclerosis Association. "What Is Amyotrophic Lateral Sclerosis? Some Questions and Answers."

Gadjusek, D. C., and Salazar, A. M. "Amyotrophic Lateral Sclerosis and Parkinsonian Syndromes in High Incidence Among the Auyu and Jakai People of West New Guinea." *Neurology* 32 (1982):107–26.

Spencer, P. S. "Guam Amyotrophic Lateral Sclerosis-Parkinsonism-Dementia Linked to a Plant Excitant Neurotoxin." *Science* 237 (1987):517–22.

Brown, D. T., and Shuster, R. "Ex-49ers Race to Unravel a Fatal Link: Three from '64 Team Contracted Rare Lou Gehrig's Disease." *USA Today*, January 22, 1987.

"Milorganite Reports Demand Precaution." *Milwaukee Sentinel*, February 11, 1987.

Behm, D. "Milorganite: Do We Have a Problem Here?" *Milwaukee Journal*, February 15, 1987.

Grogan, D. W., et al. "An Incurable Killer Strikes Three Ex-49ers, and an Anguished Victim Doubts It's a Coincidence." *People*, February 9, 1987, pp. 94–95.

Collins, T., and Manning J. "Lou Gehrig's Disease Group Backs Milorganite Study." *Milwaukee Sentinel*, February 7, 1987.

Collins, T. "Official to Urge No Study of Milorganite and ALS." *Milwaukee Sentinel*, February 12, 1987.

Hayes, P. "UW Scientist Pursues Possible ALS-Milorganite Link." *Milwaukee Journal*, March 1, 1987.

Collins, T., and Manning, J. "49er's Wife Seeks Milorganite Study." *Milwaukee Sentinel*, March 9, 1987.

Diverticulosis

Baldwin, W. M. "Duodenal Diverticulum in Man." *Anatomical Records* 5 (1911):121–25.

Case, J. T. "Roentgen Observations on the Duodenum with Special Reference to Lesions Beyond the First Portion." *American Journal of Roentgenology* 3 (1916):314–17.

Appendicitis

Fitz, R. H. "Perforating Inflammation of the Vermiform Appendix: With Special Reference to Its Early Diagnosis and Treatment." *American Journal of Medical Science* 92 (1886):321–46.

Sarcoidosis

Boeck, Caesar. "Multiple Benign Sarkoid of the Skin." *Archives of Dermatology* 118 (1982):711–20.

Lupus

Dubois, E. L. *Lupus Erythematosus*. Los Angeles: University of Southern California Press, 1976.

Dubois, E. L. "Information for Patients with Lupus Erythematosus." American Lupus Society, 1983.

Hyperactivity and Learning Disability

Committee on Nutrition of the American Academy of Pediatrics. "Should Milk Drinking by Children Be Discouraged?" *Pediatrics* 53 (1974):576–82.

Feingold, B. F. *Why Your Child Is Hyperactive*. New York: Random House, 1975.

Egger, J., et al. "Controlled Trial of Oligoantigenic Treatment in the Hyperkinetic Syndrome." *Lancet*, 1 (1985):540–45.

McCarrison, Sir Robert. *Studies in Deficiency Diseases*. New York: Oxford University Press, 1921.

McCarrison, Sir Robert. *Nutrition and Health*. London: Faber & Faber, 1922.

Banik, Allen, and Taylor, Renée. *Hunza Land*. Long Beach, Calif.: Whitehorn Publishing Co., 1967.

Eddins, S. "School Vetoes Junk Food." *Pritikin Research Foundation Newsletter* 2 (1982):1–4.

Weinstock, C. P. "Doubt Prisoners' Calm Behavior Linked to Diet Sans Sugar and Bread." *Medical Tribune*, January 30, 1985.

Allergy

Talbot, F. B. "Idiosyncrasy to Cow's Milk." *Boston Medical and Surgical Journal* 175 (1916):409.

Davies, W. "Cow's Milk Allergy in Infancy." *Archives of Diseases of Childhood* 33 (1958):265.

Egger, J., et al. "Is Migraine Food Allergy? A Double-Blind Controlled Trial of Oligoantigenic Diet Treatment." *Lancet* 2 (1983):865–69.

Burney, P. G. J. "Asthma Mortality in England and Wales: Evidence for a Further Increase, 1974–84." *Lancet* 2 (1986):323–26.

Jackson, R. T., et al. "Mortality from Asthma: A New Epidemic in New Zealand." *British Medical Journal* 285 (1982):771–74.

Sly, R. M. "Increases in Deaths from Asthma." *Annals of Allergy* 53 (1984):20–25.

Turner, K. J., et al. "Studies on Bronchial Hyperreactivity, Allergic Responsiveness, and Asthma in Rural and Urban Children of the Highlands of Papua, New Guinea." *Journal of Allergy and Clinical Immunology* 77 (1986):558–66.

Chapter 17: Primitive Versus Industrialized Societies

Balke, B., and Snow, C. "Anthropologic and Physiologic Observations on Tarahumara Endurance Runners." In *Health Problems of U.S. and North American Indian Populations*. New York: MSS Information Corp., 1972.

Norman, James. "The Tarahumaras: Mexico's Long Distance Runners." *National Geographic* 149, no. 5 (1976).

Mourstaff, G. J., et al. "Diabetes Mellitus in Eskimos." In *Health Problems of U.S. and North American Indian Populations*. New York: MSS Information Corp., 1972.

"Eskimo Diets and Diseases." *Lancet*, May 21, 1983.

Rabinowitch, I. M. "Clinical and Other Observations of Canadian Eskimos in the Eastern Arctic." *Canadian Medical Association Journal* 34 (1936):487–501.

Gottman, A. W. "A Report of 103 Autopsies on Alaskan Natives." *Archives of Pathology* 70 (1960):117–24.

Arthaud, B. "Cause of Death in 339 Alaskan Natives as Determined by Autopsy." *Archives of Pathology* 90 (1970):433–38.

Dyerberg, J., and Bang, H. O. "Lipid Metabolism, Atherogenesis, and Haemostsasis in Eskimos: The Role of the Prostaglandin-3 Family." *Haemostasis* 8, nos. 3–5 (1979):227–33.

Dyerberg, J., et al. "Eicosapentaenoic Acid and Prevention of Thrombosis and Atherosclerosis?" *Lancet* 2 (1978):117–19.

Dyerberg, J., and Bang, H. O. "Haemostatic Function and Platelet Polyunsaturated Fatty Acids in Eskimo." *Lancet* 2 (1979):433–35.

Zimmerman, M. R. "Paleopathology in Alaskan Mummies." *American Scientist* 73 (1985):20–25.

Mazess, R. "Bone Mineral Content of North Alaskan Eskimos." *American Journal of Clinical Nutrition* 27 (1974):916–20.

Biss, K., et al. "Some Unique Biological Characteristics of the Masai of East Africa." *New England Journal of Medicine* 284 (1971):694–99.

Mann, G. V., et al. "Atherosclerosis in the Masai." *American Journal of Epidemiology* 95 (1972):26–37.

Mann, G. V., et al. "Cardiovascular Disease in the Masai." *Journal of Atherosclerosis* 4 (1964):289–312.

Burstyn, P. G., and Gibney, W. J. "Milk, Serum Cholesterol and the Masai." *Atherosclerosis* 35 (1980):339–43.

Shaper, A. G. "Cardiovascular Studies in the Samburu Tribe of Northern Kenya." *American Heart Journal* 63 (1962):437–42.

Shaper, A. G., et al. "Serum Lipids in Three Nomadic Tribes in Northern Kenya." *American Journal of Clinical Nutrition* 13 (1963): 135–45.

Singer, C., et al. *A History of Technology.* Vol. 3, *From the Renaissance to the Industrial Revolution*; Vol. 4, *The Industrial Revolution*; Vol. 5, *The Late Nineteen Century.* London: Oxford University Press at the Clarendon Press, 1958.

Von Meyer, Ernst. *History of Chemistry.* New York: Macmillan & Co., 1896.

Lomax, E. R., "Advances in Pediatrics and in Infant Care in Nineteenth-Century England." Ph.D. diss., UCLA, 1972.

"Some Demographic Aspects of Aging in the United States." Washington, D.C.: National Center of Health Statistics, 1983.

Chapter 18: The Causes of the Man-Made Diseases

Cochran, R. A. "PCBs: Yesterday's Wonder Chemicals Become a Major Pollution Problem." *Analytical Control* (NUS Corporation) 7 (1982):1–4.

Steffey, K. L., et al. "A Ten-Year Study of Chlorinated Hydrocarbon Insecticide Residues in Bovine Milk in Illinois." *Journal of Environmental Science and Health, Part B: Pesticides, Food Contaminants, Agri Wastes* B19 (1984).

Lipscomb, G. Q., and Duggan, R. E. "Dietary Intake of Pesticide Chemicals in the U.S." *Pesticides Monitoring Journal* 2 (1969):162–69.

Cornelliussen, P. E. "Pesticide Residues in Total Diet Samples, IV, VI. *Pesticides Monitoring Journal* 2 (1969):140–52; 5 (1972):313–30.

Price, J. M. *Coronaries, Cholesterol, Chlorine.* New York: Pyramid Books, 1969.

Clifford, D. A. "Point-of-Use Treatment for Nitrate Removal." *Water Technology* 7, no. 8: 26–36.

Guter, G. "Removal of Nitrate from Water Supplies for Public Use." EPA Report No. 6002-81-029, April 1981.

Barbeau, A., et al. "Ecogenetics of Parkinson's Disease: 4-Hydroxylation of Debrisoquine." *Lancet* 2 (1985):1213–16.

*Davis, G. C., et al. "Chronic Parkinsonism Secondary to Intravenous Injection of Meperidine Analogues." *Psychiatric Research* 1 (1979):249–54.

*Langston, J. W., and Ballard, P. A. "Parkinson's Disease in a Chemist Working With 1-Methyl-4-phenyl-1,2,5,6-tetrahydropyridine." *New England Journal of Medicine* 309 (1983):310–13.

U.S. Department of Health and Human Services. Public Health Service. Food and Drug Administration. "All About FDA: An Orientation Handbook." July 1984.

Environmental Protection Agency. Office of Public Affairs (A-107). "EPA: Your Guide to the Environmental Protection Agency."

Environmental Protection Agency. Office of Public Affairs (A-107). "EPA: Regulating Pesticides."

Environmental Protection Agency. Office of Toxic Substances (TS-792). "EPA: The Toxic Substances Control Act."

*Environmental Protection Agency. Health Effects Branch, Criteria and Standards Division, Office of Drinking Water. "Draft Criteria Document For 1,2-Dichloroethane." January 1982.

*Letkiewicz, F., et al. "1,1-Dichloroethylene (Vinylidene Chloride): Occurrence in Drinking Water, Food, and Air." JRB Associates, under contract to Environmental Protection Agency. EPA Contract No. 68-01-6388, November 1983.

Index

FOR THE BEST IN PAPERBACKS, LOOK FOR THE

In every corner of the world, on every subject under the sun, Penguin represents quality and variety—the very best in publishing today.

For complete information about books available from Penguin—including Pelicans, Puffins, Peregrines, and Penguin Classics—and how to order them, write to us at the appropriate address below. Please note that for copyright reasons the selection of books varies from country to country.

In the United Kingdom: For a complete list of books available from Penguin in the U.K., please write to *Dept E.P., Penguin Books Ltd, Harmondsworth, Middlesex, UB7 0DA.*

In the United States: For a complete list of books available from Penguin in the U.S., please write to *Dept BA, Penguin,* Box 120, Bergenfield, New Jersey 07621-0120.

In Canada: For a complete list of books available from Penguin in Canada, please write to *Penguin Books Ltd, 2801 John Street, Markham, Ontario L3R 1B4.*

In Australia: For a complete list of books available from Penguin in Australia, please write to the *Marketing Department, Penguin Books Ltd, P.O. Box 257, Ringwood, Victoria 3134.*

In New Zealand: For a complete list of books available from Penguin in New Zealand, please write to the *Marketing Department, Penguin Books (NZ) Ltd, Private Bag, Takapuna, Auckland 9.*

In India: For a complete list of books available from Penguin, please write to *Penguin Overseas Ltd, 706 Eros Apartments, 56 Nehru Place, New Delhi, 110019.*

In Holland: For a complete list of books available from Penguin in Holland, please write to *Penguin Books Nederland B.V., Postbus 195, NL-1380AD Weesp, Netherlands.*

In Germany: For a complete list of books available from Penguin, please write to *Penguin Books Ltd, Friedrichstrasse 10-12, D-6000 Frankfurt Main I, Federal Republic of Germany.*

In Spain: For a complete list of books available from Penguin in Spain, please write to *Longman, Penguin España, Calle San Nicolas 15, E-28013 Madrid, Spain.*

In Japan: For a complete list of books available from Penguin in Japan, please write to *Longman Penguin Japan Co Ltd, Yamaguchi Building, 2-12-9 Kanda Jimbocho, Chiyoda-Ku, Tokyo 101, Japan.*